Thomas Kingsmill Abbott

The Elements of Logic

Second Edition

Thomas Kingsmill Abbott

The Elements of Logic
Second Edition

ISBN/EAN: 9783337278007

Printed in Europe, USA, Canada, Australia, Japan

Cover: Foto ©berggeist007 / pixelio.de

More available books at **www.hansebooks.com**

THE ELEMENTS OF LOGIC.

BY

T. K. ABBOTT, B.D.,

FELLOW AND TUTOR OF TRINITY COLLEGE, DUBLIN.

SECOND EDITION.

DUBLIN: HODGES, FIGGIS, & CO., GRAFTON-STREET.
LONDON: LONGMANS, GREEN, & CO.

1885.

TO THE READER.

In the following treatise I have endeavoured to unite conciseness and scientific accuracy, while adhering, as far as possible, to the traditional lines of the Aristotelian Logic. Critical discussions have been avoided, as out of place in a purely Elementary Treatise, but the student who desires to pursue the subject further will have nothing to unlearn. A few things have been included which could not with propriety have been omitted, even in an Elementary Treatise, but which may be passed over by the student who only seeks a minimum of knowledge of Logic. The paragraphs containing these are printed with close lines, and are (with insignificant exceptions) further distinguished in the Table of Contents by a prefixed [†].

CONTENTS.

N.B.—*Paragraphs printed closely, and marked* [†] *in this Table, may be passed over by the beginner.*

PART FIRST.

OF TERMS.

	PAGE
Preliminary,	3
Of Terms in General,	3
Abstract of Concrete Terms,	4
Common, Singular, and Collective Terms,	5
Of Denotation and Connotation,	5
Of Abstraction and of Genus and Species,	7
[†] Remarks on Concepts,	8
Non-connotative Terms,	8
Contradictory and Contrary Terms,	10
Clearness and Distinctness,	10
Of Definition,	11
Of Division,	12

PART SECOND.
OF SIMPLE PROPOSITIONS.

	PAGE
Preliminary Definitions,	14
The Copula,	15
Quality of Propositions,	16
Quantity of Propositions,	16
Distribution of Terms,	18
[†] Quantification of the Predicate,	20
Of the Import of Propositions,	21
Verbal and Real, or Analytical and Synthetical Propositions,	23
Copulative Propositions,	24
Modal Propositions,	25

APPENDIX TO PART II.

[†] Of the Predicables,	26
[†] Of the Ten Categories,	27

PART THIRD.
OF INFERENCES.

CHAPTER I.
Of Immediate Inferences.

Subalternation,	29
Opposition,	30
Conversion,	34
Contraposition,	36
Remarks on Conversion,	38
Of the Principle of Substitution,	39

CONTENTS.

CHAPTER II.

Of Mediate Reasoning, or Syllogism.

	PAGE
Of Mediate Reasoning Generally,	40
General Rules of Syllogism,	42
[†] Inferences from premisses with Undistributed Middle,	44
Of the Four Figures,	44
Special Rules of the Four Figures,	47
[†] Remark on the Special use of each Figure,	50
Of Moods,	51
[†] Remark on the Validity of the Moods enumerated,	53
Of Aristotle's Dictum,	55
Of Reduction,	57
[†] Of Reductio ad Impossibile,	61
Of the Unfigured Syllogism,	62
Of the Enthymeme,	63
Of Sorites,	63

CHAPTER III.

Of Complex Propositions and Syllogisms.

Of Complex Propositions,	65
Of Complex Syllogisms,	66
Conditional Syllogisms,	67
Disjunctive Syllogisms,	68
Of the Dilemma,	69
Of the Reduction of Complex Syllogisms,	70

CHAPTER IV.

Of Probable Reasoning.

Of Chains of Probability,
Of Cumulative Probabilities,
[†] Of Inductive Reasoning,
[†] Of the **Logical** Basis of Induction,
[†] **Mr.** Mill's View of the Type of Reasoning,
Of Analogy,

PART FOURTH.
CHAPTER I.
Of Fallacies.

Of Fallacies in General,
Of Logical Fallacies,
   Illicit Process,
   Undistributed Middle,
   Two Middle Terms,
Fallacies in Complex Syllogisms,
Of Semi-logical Fallacies,
   Ambiguous **Terms,**
   Composition and Division.
   Fallacy of Accident,
Of Material Fallacies,
   Ignoratio **Elenchi,**
   Petitio Principii,
   Arguing in **a** Circle,
   A non causa pro causa,
   Fallacy of Many Questions,
   Fallacy of False Analogy,
   Argumentum ad Hominem,

CHAPTER II.

Of Methods of Proof and of Exposition.

	PAGE
Analytic and **Synthetic Exposition**, ...	90
A priori and **A posteriori Proof**, ...	91
Of **Explanations** and **Deduction**, ...	91

APPENDIX.

EXERCISES, ...	93
INDEX, ...	101

CORRIGENDA.

Page 52, line 3.—*After* premisses *add* I E O, however, is illegitimate, for **the** term is particular in its premiss and universal in **the co**sion.

Page 52, line 3.—*For* twelve *read* **eleven**.

Page 52, line 4.—*For* eleven *read* ten.

THE ELEMENTS OF LOGIC.

INTRODUCTION.

WHAT IS LOGIC?

1. **EVERYTHING** in nature takes place according to laws. This, which is true of material things, is also true of the operations of the mind. These follow laws of which we are not always aware. Language, for instance, follows laws of grammar, and these laws in any particular language are the same, whatever the subject may be of which we are speaking. The exercise of the understanding in particular is itself governed by laws. Some of these only apply to special kinds of subject-matter, as, for example, the laws of number in mathematical reasoning. But there are others which hold good in all cases, whatever the subject-matter of our thoughts or reasonings may be. It is of these laws that Logic treats. Just as the rules of grammar are the same whatever the subject spoken of may be, so the rules of Logic are the same, whatever the subject thought about or reasoned about.

2. The distinction involved in this statement is the important one between **Form** and **Matter**. Matter is the **variable** part; **Form** the **invariable**. In language, for example, the lexicon deals with the Matter; grammar deals with the Form.

3. **Logic**, then, is the Science of the **Form** of **Thought**.

4. Some writers assign to Logic a much wider sphere than this; they regard it as including the rules of estimation of evidence, and of the operations of the understanding subsidiary thereto. The narrower science here treated of may, for the sake of precision, be called **Formal Logic**, and the exposition of the rules just referred to may be called **Applied Logic**.

PART FIRST.

OF TERMS.

Preliminary.

5. THE simplest act of thought is a Judgment, which expressed in words is a Proposition. Propositions so connected that one follows from one or more others constitute Reasoning. Propositions, however, are compound in expression, and consist of words or terms. Hence we begin with Terms; we treat in the next place of Propositions, and in the third place of Reasoning.

Of Terms in General.

6. In a proposition we either affirm or deny something of something else. 'Horses are quadrupeds;' 'horses are useful animals.' Here we affirm 'quadrupeds,' 'useful animals,' of 'horses.' Words which can be used in this way, *i. e.* as naming something which we affirm or deny, or as naming that of which something is affirmed or denied, are called Terms. That which is affirmed or denied is called the Predicate; that of which it is affirmed or denied is the Subject.

7. A Term, therefore, may be defined as a word or combination of words, which can be used by itself as subject or predicate of a proposition. It must be observed that a Term may consist of several words. In

the proposition given above, 'useful animal,' is a single term; 'Queen of Great Britain and Ireland' is also a single term.

8. On the other hand, there are many words which cannot stand by themselves as terms. Such are prepositions, adverbs (except of time or place), conjunctions, oblique cases of nouns, &c. These can be used as terms only with something understood. We might say: '"of" is a preposition,' *i.e.* the word or sound 'of.'

'Term' is derived from 'terminus,' used by logicians to represent the Greek ὅρος; the subject and predicate being, as it were, the boundaries of the proposition.

Of Abstract and Concrete Terms.

9. Terms may be names of Things, or of Attributes of Things. For example, 'man' is a name of Things; 'humanity' is the name of an Attribute. **Names of Things** are concrete; **Names of Attributes are Abstract.** Thus 'whiteness' is Abstract; 'white thing' is Concrete.

10. Concrete Terms may be grammatically either substantives or adjectives (including participles). 'This horse is white' means the same as 'this horse is a white horse' or 'a white thing.' An adjective cannot always be used alone as the subject of a proposition; this is because it is, grammatically, an incomplete name, and requires a substantive either expressed or understod to complete it. When used as a predicate, the subject is understood; as in the example just given—'This horse is white' (*i.e.* 'a white horse'). Greek and Latin are more free in their use of adjectives.

The same word may be sometimes used as an abstract,

and sometimes as a concrete term, *ex. gr.* 'green is an agreeable colour.' Here 'green' = 'green colour.' *

Of Common, Singular, and Collective Terms.

11. Names of things are divided into **Singular**, **Common**, and **Collective**.

12. A **Singular Term** is one which names an individual thing, as: Socrates, the present Prime Minister, this person, the oldest inhabitant. A **Common** or **Universal Term** is applicable to any one of a class, which may include many things, as: man, tree, phoenix, prime minister.

13. A **Collective Term** is applicable to a group of things as a whole, but not to each member of it, as: the British Parliament, the fifty-third regiment.

14. A singular term cannot be a predicate unless the subject is also a singular term. There are some abstract terms which are names of classes of attributes, as: colour, sound.

Of Denotation (or Extension) and Connotation.

15. Universal terms have a double signification. By the word 'boat,' for example, we mean a small open vessel constructed to float on the water, and to carry one or more persons, etc. But if asked of what is the word 'boat' a name, we might answer either by giving this definition, or by enumerating different kinds of boats, as barges, pinnaces, gigs, etc., or all individual boats. The

---

* 'Abstract' is sometimes used, especially by the older writers, in the sense of 'general.'

term 'boat' is said to 'denote' all these, that is, everything which can be called a boat, and to 'connote' or note therewith the attributes implied in the name.

16. The things to which the name is applicable constitute its Denotation, or, as it is sometimes called, its Extension. **The attributes implied by the name constitute its Connotation or** comprehension. The answer to the question, What is a boat, *i. e.* what is the meaning of the word, is its Connotation. The answer to the question, Of what things can we say that they are boats, is its Denotation or Extension. The form of the question may be such as to admit of either answer. 'What are boats?' 'What are the metallic elements?' 'Who are the greatest men in history?' may be understood in either way.

17. It is obvious that the larger the number of different things included in the Extension (= Denotation) the fewer will be the common attributes; and, on the other hand, the greater the number of attributes taken into the Connotation (= Comprehension) the fewer, in general, will be the individual things to which the term is applicable. In other words: By increasing the Connotation we diminish the Extension, and by enlarging the Extension we diminish the Connotation. Thus, if we take the term 'house,' and add the attribute 'intended for dwelling in,' we get 'dwelling-house,' which includes (or denotes) fewer objects, that is, has a less Extension. On the other hand, if we wish to use a term which will apply to (or denote) a greater number of things, such as 'structure,' we must necessarily leave out of the Connotation those attributes which distinguish houses from other structures.

18. It may happen in particular cases that we may add

to the Connotation of a term without increasing the extension. For instance, we may add to 'rational animal' the attribute 'biped;' and since all rational animals are bipeds, the extension remains unaltered.

Of Abstraction, and of Genus and Species.

19. If we take any term whose connotation is known, and leave out some of the attributes connoted, we thereby form a new notion which denotes, or may denote, a larger number of individuals. It is called a more general notion. This leaving out of attributes is called 'abstracting' from them, and the process itself is 'Abstraction.'

20. The notion thus arrived at is called the 'Genus' of the less general; the less general is the 'Species.' When two common terms are so related that the connotation of one includes the connotation of the other, that which has the larger connotation is called the Species of the other, which is called the Genus.

21. Thus, if we take the term 'riding horse,' and leave out the notion implied in the first part, we arrive at the notion 'horse,' which is the Genus, of which 'riding horse' is Species. On the other hand, if we add to the latter the notion 'small,' we arrive at 'riding pony,' which is a Species, of which 'riding horse' is Genus.*

22. The more general notion is said to be higher than the less general, which is lower.

23. It is clear that whatever attribute belongs to the

---

* N. B.—The terms 'Genus' and 'Species' have a more limited and technical sense in treatises on Natural History.

higher notion **belongs also to** the lower, **and** whatever is inconsistent with **the higher is** inconsistent **with** the lower. For the notion of **the lower includes or** contains in it the notion of the **higher. On the** other hand, the extension of the lower **is** part of the extension **of** the higher, and, therefore, whatever individual belongs **to the lower** class belongs also **to** the higher.

24. A Genus which **has** none above it is a **Genus summum**; a Species **which** has none below it is a **Species infima.** A Summum Genus with its subordinate genera, and these again subdivided into lower genera, and so on to the **lowest** species, is **a Predicamental line.**

Remarks.

25. These universal terms may **be** regarded as names **of** notions or concepts, the concept being the group of attributes which constitutes the meaning of the term.

26. In arriving at concepts **or** universal terms, we generally begin by grouping together things **which** have a certain general resemblance not yet definitely stated, *i. e.* we begin by using the term to designate **a** certain denotation. We **may** afterwards proceed **to study the** common attributes **implied** in the term, and we may find that in our application **of** the term we have gradually dropped out of sight **one** attribute after another until there is little in common between the things **first** and those last named. Thus '**boat**' is applied to a steam-packet, **as well** as to a row-boat.

Of Non-Connotative Terms.

27. Some **terms** have no connotation, *e. g.* **proper** names. The fact **that the name** 'Walter Scott' calls up to our mind the idea **of** the 'Author of Waverley' does not prove that the name connotes **or** implies this. It was not because he wrote 'Waverley' that he was called Walter Scott; **and if it** were proved that **he** had **not written it**

he would not cease to be so called, nor would the actual author acquire a right to the name. The name simply denotes a person living in a certain time and place (unless we say that it connotes the male sex). It may be asked: 'If the name John Smith does not suggest a certain appearance, &c., how am I to know him when I see him?' The answer is, it may *suggest*, but that is not the same as connoting, any more than the cart is part of the horse that draws it. The name 'earth-worm' may suggest 'Charles Darwin' (because he wrote a treatise about earth-worms), but it does not connote it. What is decisive is, that if another man were found very like my mental picture of John Smith, I should not call him by that name. So the name 'London' calls up to my mind the notion 'the largest and richest city in the world.' But this is not connoted by the name. The city was called London when it was not the largest and richest; it would continue to be so called although we should discover that Pekin was larger and richer, and we should not in that case call the latter city London.

28. A term which, though not a proper name, applies only to a single individual, may be connotative. Thus, 'the Author of Waverley' denotes an individual and connotes his attribute. It is not, however, a proper name, only we happen to know that it can apply only to one individual.

29. The name of a single attribute is also non-connotative; a name like 'colour,' which applies to several attributes, connotes that in which they agree; in this case, causing a certain kind of visual sensation.

Of Contradictory and Contrary Terms.

30. Terms or attributes, of which one is simply the negative of the other, are called **Contradictory**, as: wise, not wise.

31. Contradictory Attributes cannot at the same time belong to the same object. This is called the **Law of Contradiction**.

32. Of two Contradictory Attributes one or other must be predicable of every object. This is called the **Law of Excluded Middle**; middle having here the sense of *mean*. Between 'white' and 'not white' there is no mean.

33. Terms or attributes which are the most opposed of those coming under the same class are called **Contrary**, as: wise, foolish; white, black.

34. Contrary terms cannot at the same time be predicated of the same object.

Of Clearness and Distinctness.

35. A notion, or concept, or meaning of a term, is said to be **clear** when we are able to distinguish it from other notions or concepts. It is **distinct** when we are able to distinguish and enumerate the parts (attributes) of which it is composed.

36. A notion may be tolerably clear, and yet far from being distinct. Thus, we sometimes say we know very well what the thing is, but cannot define it, *i. e.* we know of what things it may be predicated. This is the case with natural objects generally, and with many other objects of experience. We may have a sufficiently clear notion of horses and of trees, but might find it impossible to analyze

either notion with any correctness. We may know what is poetry and what is not, but should find it hard to define what constitutes poetry.

37. The opposite of clear is **obscure**. The opposite of distinct is **confused**.

Of Definition.

38. **Definition** is a statement of the connotation of a term.

39. The rules of Definition are:—

(*a*) It must be adequate (*adæquata*), that is, its extension must be exactly equal to that of the term defined. If this rule is violated the definition is either too wide or too narrow.

40. (*b*) It must be precise, that is, it must contain nothing superfluous. Thus it would be incorrect to define a parallelogram as a rectilinear quadrilateral whose opposite sides are equal and parallel; for the equality follows from the parallelism. Such a definition would imply that we could have a rectilinear quadrilateral whose opposite sides were parallel and not equal, or equal and not parallel.

This rule might be otherwise expressed by saying that the attributes enumerated must be **distinct**.

41. (*c*) It must not contain any term equivalent to the term defined; for instance, we must not define Life as the Operation of Vital Forces, for 'vital' means 'of life.'

The violation of this rule constitutes a 'circle in definition.'

42. (*d*) *If possible*, it should not be by negative attributes.

43. It is to be observed that definitions of natural objects or kinds of things, such as we know by experience, can seldom be complete. We can only define the name so as to mark out clearly the thing intended. Natural kinds of things possess an indefinite number of attributes in common, and it is to a certain extent arbitrary which of these we select as the connotation of the name. Even after we have made this selection, the discovery of a thing differing in one or two of these attributes might induce us to alter the connotation.

44. The definitions actually adopted by naturalists, botanists, &c., are not based on the ordinary connotation of the names, but on a selection of attributes supposed to be more fixed or connected with a greater number of other attributes.

45. **Description** is an enumeration of attributes of the thing described sufficient to distinguish it from other things.

46. Individual things can only be described, not defined.

Of Division.

47. **Division** is an enumeration of the parts of the Extension or Denotation of a Term. Thus Animal is divided into Vertebrate and Invertebrate.

48. The chief rules of Division are:—

(*a*) It must be adequate, that is, all the members taken together must be equal to the whole.

49. (*b*) It must be distinct; that is, the members must be mutually exclusive; in other words, each must be capable of being denied universally of the rest. For instance,

we must not divide European into French, German, Prussian, &c.

50. (*c*) *If possible*, there should be only **one Principle of Division** (*fundamentum divisionis*), that is, all the dividing attributes should come under one motion. For instance, we might divide Book according to contents into histories, novels, &c., or according to size into folios, quartos, &c., or according to binding into bound and unbound; but it would be incorrect to divide Book into folios, quartos, novels, &c. If the third rule is observed the second will be also fulfilled, but it is not always possible to observe this third rule.

51. Individuals are so called as not being capable of logical division.

PART SECOND.

OF SIMPLE PROPOSITIONS.

Preliminary Definitions.

52. A **Simple Proposition** asserts or denies one term of another term.

>Gold is heavy;
>Some planets are bright;
>No planets are self-luminous.

53. We must repeat some definitions:

The term of which something is affirmed or denied is the **Subject**.

That which is affirmed or denied is the **Predicate**.

The Predicate is said to be predicated affirmatively or negatively of the subject.

The connecting word (is; is not) is the **Copula**.

54. The type of the simple proposition is—

>A is B; or, A is not B.

A is the Subject; B, the Predicate; 'is' or 'is not,' the Copula.

55. If we use the symbol A B to signify A that is B, we may write the proposition Every A is B, thus :

> Every A is AB.

Thus if A = crow and B = black, the proposition :

> All crows are black

becomes

> All crows are black crows.

Of the Copula.

56. The only Copula admitted by logicians is the present tense of the verb 'to be.' The following remarks must be attended to :—

57. (*a*) All other verbal forms combine predicate and copula, and for logical purposes must be resolved into these two elements. Thus :

> All planets revolve round the sun

is equivalent to

> All planets are revolving (or bodies that revolve) round the sun.

58. (*b*) The Copula does not involve the notion of time, although it has the grammatical form of the present tense. If the assertion refers specially to past, present, or future time, the note of time must be put in the predicate. Thus: Every man was a boy = Every man is what was formerly a boy. Babylon was once the capital of a great empire = Babylon is a city once the capital, &c.

59. (*c*) The Copula does not involve the notion of existence. Thus: Centaurs are imaginary beings, half man, half horse, is a proposition logically correct.

Of the *Quality of Propositions.*

60. By the **Quality** of a proposition (*qualitas*) is meant the answer to the question : Of what sort (*qualis*) is the predication ? *i. e.* is it **Affirmative** or **Negative** ?

As to **Quality**, then, propositions are divided into **Affirmative** and **Negative**.

61. Every negative proposition may be treated as an affirmative by affixing the particle 'not' to the predicate instead of to the copula—

A is not B = A is – not-B ;

Ex. gr. Ostriches do not fly
= Ostriches are creatures that do not fly.

62. Similarly, Affirmatives may be changed into Negatives. Thus :

Worms are invertebrate
= Worms are not vertebrate.

These are called **Equipollent** Propositions.

Of the *Quantity of Propositions.*

63. The **Quantity** of a Proposition is the answer to the question : What is the **extent of the assertion ?**

In respect to **Quantity**, then, Propositions are divided into **Universal, Particular**, and **Singular**.

64. A **Universal Proposition** is one in which the **Subject is a** common term **taken in its whole extension :**

Every crow is black, *or* All crows are black ;
No reptiles fly.

In symbols,

Every A is B ;
No A is B.

65. A **Particular Proposition** is one in which the Subject **is a common term taken in part of its extension**:

>Some swans are black;
>Some swans are not black.

In symbols,

>Some As are B;
>Some As are not B.

66. It must be observed that 'some' in logic means 'one at least,' and does not mean 'some only.' 'Some As are B' is true if one A is B and if every A is B.

67. A **Singular Proposition** is one in which the **subject is a Singular, Collective, or Abstract** term, as—

>John is tall;
>The English nation is of mixed origin;
>Truthfulness is a virtue.

Cases where a term originally abstract is used as a common term are of course not included.

68. For logical purposes Singular Propositions may generally be regarded as universal, the subject being taken in its whole extension (which is only one thing).

69. It is possible, however, to make a Proposition which has a singular term as its subject, particular, by distinguishing different times of the individual's existence. *Ex. gr.* 'John is not always busy.' Here the real subject may be considered to be 'the times of John's existence,' and the assertion is, that some of these times are not busy times. So with abstract terms: 'Law is not always justice' is equivalent to 'Some things legal are not just.' 'Justice is ever equal' is equivalent to 'All justice is equal.'

70. Propositions are sometimes expressed indefinitely, as: Men are cruel to men. But before we can deal with such a proposition the subject must be 'quantified,' that is, it must be specified whether the assertion is to be understood of all men or of some. This can be done only by reference to the 'matter;' and not by any rules of logic. In the absence of further information we must assume that such propositions are particular.

71. Since propositions are twofold as to quality and twofold as to quantity, we have four kinds of propositions, and these are symbolically represented as follows:—

    A universal affirmative is called **A**.
    A universal negative   ,,   **E**.
    A particular affirmative ,,   **I**.
    A particular negative   ,,   **O**.

(The vowels A and I were chosen as the first two of 'affirmo;' E and O as the vowels of 'nego.')

72. It must be observed that by the idiom of the English language 'Every A is not B,' or 'All As are not B,' means 'Not every A is B,' *i.e.* 'Some As are not B.'

Of the Distribution of the Terms of a Proposition.

73. A common term is said to be distributed when it is taken in its whole extension; otherwise it is undistributed.

74. The quantity of a proposition has been already defined to be the quantity of its subject, *i.e.* a universal proposition is one whose subject is distributed, and a particular proposition is one whose subject is undistributed. The quantity of the subject and the quantity of the proposition are therefore one and the same.

75. The quantity of the predicate depends on the quality of the proposition.

> The predicate of an affirmative proposition is undistributed (= is particular).

> The predicate of a negative proposition is distributed (= is universal).

76. First, when we say 'Crows (all or some) are black,' we do not imply that other things may not be black also. Hence the predicate 'black' is not taken in its whole extension. We merely assert that crows are reckoned amongst black things; *i.e.* the extension of 'crow' is part (at least) of the extension of 'black.' The same may be said of every affirmative proposition. 'A is B' does not mean that A is the only thing that is B.

77. Hence the predicate of every affirmative proposition is particular (= is undistributed). It may happen that the extension of the predicate does not exceed that of the subject, as when we say 'All equilateral triangles are equiangular,' and it is also true that all equiangular triangles are equilateral. But this is accidental, and was not contained in the former proposition.

78. Secondly, when I say 'No crows are white' I assert that no crow is any white thing; *i.e.* no white thing is identified with a crow. Therefore 'white' is taken in its whole extension.

79. So if the proposition is a particular negative: 'Some swans are not white,' *i.e.* are not any white thing, I assert that some swans are excluded from the whole extension of white things.

80. Hence the predicate of every negative proposition is distributed (= is universal).

81. Using the symbols **introduced** in the last section:—

A has its subject universal and predicate particular.

E has both subject and predicate **universal.**

I has both subject and predicate **particular.**

O has its subject **particular and predicate universal.**

Of the Quantification of the Predicate.

82. Some logicians maintain that in every proposition we in thought assign a definite quantity to the predicate, and therefore ought to do so in words when we wish to make the proposition logically complete. They hold that when we say 'All crows are black,' we mean 'All crows are some black things;' and when we say 'Equilateral triangles are equiangular,' we mean '**All** equilateral triangles **are all** equiangular triangles.'

83. On this **view,** then, affirmative propositions may have **a** universal predicate, and negative propositions may **have** a particular predicate, as: (All or some) A is not some (certain) B. This is called 'quantifying' the predicate.

84. Every affirmative proposition **is** on this doctrine viewed as an equation **in** extension; every negative **as an** inequality; and there **are,** of course, eight forms of propositions, instead **of four, A E I O** being each subdivided into two, according **to the** quantity of the predicate.

85. It **deserves to** be noticed that in these forms the words 'all,' 'some,' are used in a modified sense. 'All' **does not** mean 'every,' but 'all together,' *i.e.* it is not taken distributively, but collectively. '**Some**' is also taken collectively and definitely, as '**some** certain' (= 'quidam').

86. In **common** language we sometimes quantify the predicate; *ex. gr.* by using such words **as** 'alone,' 'only,' 'constitute,' 'consist of,' &c. **Thus:**

The virtuous alone are truly happy = A alone is B = A is all B.

Dead languages are not the only ones worth studying = A is not some B.

A caustic is a kind of curve = A is some B.

There are patriots and patriots = There are different kinds of patriots = Some A is not some A.

87. Some of these propositions would be regarded on the common theory as compound. For instance, 'All A is all B' would be regarded as including two assertions: 'All A is B' and 'All B is A.' 'The virtuous alone are truly happy' means 'The virtuous are happy, and the not virtuous are not happy.'

Of the Import of Propositions.

88. Although affirmative propositions have all been reduced to the type (Every or Some) A is B, it does not follow that the relation between A and B is in all cases the same.

The distinction most important to notice is, that **some propositions express complete coincidence or equality; others partial** coincidence.

89. Coincidence is expressed in the following cases:—

(*a*) When the subject and predicate are both singular, both collective, or both abstract terms (used in their proper sense, *i.e.* not figuratively). Both singular, as: Victoria is Queen of Great Britain and Ireland. Both collective, as: The House of Lords is the Hereditary Chamber. Both abstract, in which case they must be synonymous: Freedom is Liberty.

(*b*) When the predicate is a definition: Water is H$_2$O.

(*c*) In mathematical propositions of quantity: The square of four is sixteen: The line A is equal to the line B; *i.e.* The magnitude of A is the magnitude of B.

90. In other cases what is implied in affirmative propositions is partial coincidence: **the attributes** connoted by **the** predicate **are** asserted to **co-exist** with those connoted by the subject, and hence the extension of the subject is contained under the extension of the predicate; **in** other words, coincides with part of it, *ex. gr.* Whales are mammalia; Whales are of prodigious size.

91. Where the predicate is an adjective, the proposition usually asserts merely the possession **of** an attribute: where it is a substantive, the proposition asserts that the subject belongs **to a** certain class. But this is only an accident of language, and does not necessitate **any** logical distinction.

92. In general, therefore, **in** affirmative propositions, **the extension of the subject coincides with part, or the whole, of the extension of the predicate.**

93. N. B.—If A and B each **connote** only one attribute, or definite group **of** attributes, then what is expressed is usually not mere co-existence, but dependence; *i. e.* the attribute connoted **by** B depends on that connoted by A; *ex. gr.* :

> The diligent is always successful (= diligence causes success).
> The virtuous is happy (= virtue brings happiness).

93. This relation is often expressed by using the abstract names **of** the attributes **as subject and** predicate; *ex. gr.* :

> Knowledge **is power**;
> Virtue is happiness;
> Honesty **is** the **best** policy.

94. 'Must be' is often **used** to express the dependence

of the connotation of the subject on that of the predicate as a condition:

> A historian must be laborious;
> This patient (*i.e.* a patient with such symptoms) must have been exposed to infection.

95. Propositions expressing Real Existence are of sufficient importance to be classified separately in any complete analysis of the import of propositions. There are, however, but few of them that cannot be reduced to other heads. They are such as—God exists; the world exists; the soul exists.

Of Verbal and Real, or Analytical and Synthetical, Propositions.

96. **An Analytical or Verbal Proposition is one whose predicate is already contained in the connotation of the subject** (as part or the whole of it), as: iron is a metal; body is extended; a triangle is a three-sided figure. These only state explicitly what is involved implicitly in the idea of the subject. They are called analytical because they analyze the idea of the subject; and verbal because they may be regarded as being only about the meaning of words.

97. **A Synthetical or Real Proposition is one whose predicate is not contained in the connotation of the subject**; *ex. gr.* iron is magnetic; all bodies are heavy; a triangle has its three angles together equal to two right angles. These propositions join on something to the connotation of the subject, and hence they are called synthetical. They are called real, as stating facts about things, not about words only.

98. It must be observed that the circumstance that these propositions are, perhaps, familiar to those who know what the subject is, *ex. gr.* as in the above instances, iron, triangle, &c., does not constitute them analytical. For the statement, 'iron is magnetic,' is a statement of a fact in nature which could not be discovered by analyzing our ideas. It asserts that whatever possesses the other qualities connoted by 'iron' possesses this attribute also. There is, however, a certain latitude or indefiniteness as to the qualities which we shall select to be connoted by the term 'iron,' or any other name of a kind of things; and there is implied in all propositions concerning such kinds an assertion that a thing possessing such attributes really exists.

99. All statements of which the subject is a proper name are synthetical, as appears from what has been said about such terms, viz. that they have no connotation. 'London is a great city' is a statement of fact, and could not be elicited from an analysis of the meaning of the word 'London.'

Of Copulative Propositions.

100. Copulative propositions consist of several propositions asserted together. For instance: *a.* A is both B and C; *b.* A is neither B nor C; *c.* Neither A is B nor C is D; *d.* A is B, but C is D.

These are resolved into separate propositions connected by a copulative particle: *a.* A is B and A is C; *b.* A is not B and A is not C; *c.* A is not B and C is not D; *d.* A is B and C is D.

101. Where two affirmative propositions are joined by

an adversative particle (although, but, yet, &c.), some opposition is implied; for example, 'This book is learned but interesting.' Here it is implied that most learned books are not interesting. Again, 'This task is difficult but necessary.' Here the implied opposition is not between difficulty and necessity, but between the understood inferences from them, viz., that **the thing** should be done, and **that the thing** should not be done. The implied proposition is then, 'What is difficult is usually to be avoided,' or the like. In general the adversative particle implies that the propositions **connected are not expected to** be true together.

102. When the first proposition is negative, the opposition is between its predicate and that of the affirmative, as, ' He is not industrious but idle,' so that the two propositions may even be synonymous.

103. An affirmative and a negative assertion may be united in what is apparently a single statement, *ex. gr.* Few men are born poets = Some **men** are born poets, and Most **men are not** born poets.

The remarks in this section belong more properly to grammar than to logic.

Of Modal Propositions.

104. Modal propositions are those in which the assertion is qualified by such words as 'possible,' 'impossible,' 'probable,' 'certain,' 'necessary,' 'contingent,' 'may be,' 'must be.'

105. Of these, 'certain,' 'impossible,' 'necessary,' 'must be,' are usually only indications that the proposition is universal.

'May,' when the subject is a common term, indicates that the proposition is not universal, *ex. gr.* The best devised plans may fail.

106. In cases not capable of this explanation the word expressing the mode should be made the predicate, and the proposition the subject, *ex. gr.* The harvest will probably be good = That the harvest will be good is probable. These propositions, therefore, are no exceptions to the rule that the copula can only be 'is' or 'is not.'

Appendix to Part II.

Of the Predicables.

107. The notion of the predicate may be related to that of the subject in one of four ways, viz., as: **Genus, Definition, Property,** or **Accident**.

108. We have already divided propositions into verbal (in which the connotation of the predicate is contained in that of the subject) and real (in which it is not so contained). In the former case the connotation of the predicate may be either the whole connotation of the subject or part of it. If the whole, the predicate is a **Definition**; if part only, it is a **Genus**.

109. When the connotation of the predicate is not contained in that of the subject, its extension or denotation may be the same as that of the subject, or not the same. If the extensions are the same, the predicate is a **Property** (*proprium*), *i. e.* an attribute which belongs to the subject alone: if they are not the same, the predicate is an **Accident**.

110. This is Aristotle's doctrine. Later logicians, influenced by certain metaphysical theories, enumerated five predicables: Genus, Species, **Difference**, Property, and Accident. They regarded genera and species as having their essences fixed by nature. The essence of the species was expressed by the definition; and that which was in the essence of the species over and above the essence of the genus they called the Difference, or Essential Difference. As 'genus' is now understood, there is no logical distinction between it and 'difference'; each is part of the connotation of the subject.

Of the Ten Categories.

111. By Category Aristotle meant predication. The Greek word κατηγορεῖν meant (in logic) 'to predicate.' The Ten Categories were a classification of predicates of an individual thing, according to their own signification (not in their relation to the idea of the subject as the five predicables). We can say of a thing, what it is, or what sort, or how great, or when, where, &c. These are all different ways of filling up the blank in, 'This is ———.' Aristotle's enumeration of these possible predicates were as follows:

112. 1. Οὐσία.—The 'whatness' of a thing, or what it is named, called by Latin logicians 'substantia,' 'substance,' as man, horse.
2. πόσον.—Quantity, as 'Two feet high.'
3. ποῖον (of what sort).—Quality, as 'white.'
4. πρός τι.—Relation (more correctly Comparison), as 'double,' 'greater,' 'less.'
5. ποιεῖν.—Action, as 'cuts,' 'burns.'
6. πάσχειν.—Passion, as 'is being cut or burned.'
7. ποῦ.—Ubi, as 'here.'
8. πότε.—Quando, as 'to-day,' 'yesterday.'
9. κεῖσθαι.—Situs, as 'is sitting,' 'is standing.'
10. ἔχειν.—Habitus, as 'is armed.' 'Habitus' was an incorrect Latin imitation of the Greek ἔχειν, which, used intransitively, meant to be in a certain state.

113. These may be arranged thus :—The predicate may be—*a*, a substantive ; *b*, an adjective ; *c*, an adverb ; *d*, a verb. If a substantive, it is a name of the kind of thing ; if an adjective, it may be one of quantity, quality, or comparison ; if an adverb, it may be of time or place (no others are admissible as predicates) ; if a verb, it may be active, passive, or neuter, or may express the result of an action (as the Greek perfect passive).*

---

* The student may naturally ask how it is that adverbs of place and time are admitted as predicates when no others can be used as such. The explanation is that these adverbs imply existence. 'Is here and now' means, 'Is existing here and now.' What is predicated in such cases is, therefore, Real Existence.

PART THIRD.

OF INFERENCES.

CHAPTER I.

Of Immediate Inferences.

114. Inference is the process by which one proposition is derived from one or more others.

Immediate Inference is the deduction of one proposition from another, which virtually contains it.

115. The kinds of Immediate Inference are: Subalternation, Opposition, and Conversion.

Of Subalternation.

116. **Inference by Subalternation is the inference of the Particular from the Universal.** *Ab universali ad particulare valet consequentia.*

From the proposition, All men are mortal, we may infer, Some men are mortal; This man is mortal. The particular is called the Subalterna; the universal is the Subalternans.

117. The Subalterna of A is I; that of E is O. The rules of Subalternation are:—

(1). From the truth of the universal we may infer the truth of the particular.

(2). From the truth of the particular we cannot infer the truth of the universal. *A particulari ad universale non valet consequentia*, i. e. from a *term* taken particularly we can infer nothing as to the same term taken universally.

(3). From the falsehood of the particular we may infer the falsehood of the universal. This is a consequence of the first rule.

(4). From the falsehood of the universal we cannot infer the falsehood of the **particular**. This is a consequence of the second rule.

118. From the proposition, **Every A is B**, I can infer, **Some A is B**; and conversely, if it is false that **Some A is B**, it must be false that **All A is B**. But from Some A is B we cannot infer that Every A is B; and conversely, if it is false that every A is B, it does not follow that Some A is B is also false.

Of Opposition.

119. **Inference by Opposition is the inference from the truth or falsity of one proposition to the falsity or truth of another** having the same terms **as subject and predicate, but differing in quality.**

120. There are three cases:—**Contradictory Opposition or Contradiction is between propositions each of which asserts only what the other denies.**

121. If the subject is a singular term, the contradictories differ only in quality, as:

        James is clever,
        James is not clever;
        James is standing,
        James is not standing.

Here the predicates are contradictory, for we may regard 'not' as belonging to the predicate; therefore, these come directly under the Law of Contradiction. This is the type of Contradictory Opposition.

122. If the subject is a Common term, the Contradictory propositions differ both in quantity and quality. 'Some men are black' would not be contradicted by 'Some men are not black'; for the 'some men' may be different, and then the propositions would not even be inconsistent.

Now, since the 'some men' are not specified, it is clear that, in order that 'black' may be affirmed and denied of the same man, we must deny 'black' of every man. Hence the contradictory of, 'Some men are black' is, 'No men are black.'

123. Again, if we have a universal, 'Every man is black,' the contradictory of this is not, 'No man is black'; for this not only denies the former, but denies much more. In fact these two propositions together assert and deny 'black' of each separate individual, and include, therefore, as many distinct contradictions. In order that we may assert no more than is necessary for the denial of the given proposition, it is only necessary to state that there is an exception to it; *i. e.* 'Some men (one at least) are not black.'

124. Hence contradictory propositions are—A and O, or E and I.

N.B.—When a proposition with a singular subject is qualified by 'always' or the like, its contradictory must have some word equivalent to 'not always'; *ex. gr.:*

        John is always busy;
        John is sometimes not busy.

125. Of **contradictories**, both **cannot be** true, and **both cannot be false**; hence the rule of inference is:—

From **the truth** of either contradictory we may **infer** the falsity of the other.

And *vice versâ*:

From **the** falsity of **either** contradictory we **may infer** the truth of the other.

126. Contrary Opposition is between propositions the most remote of those having the same subject and the same predicate; *i.e.* between **a** Universal Negative and a Universal Affirmative; *ex. gr.*:

       All men are black;
       No men are black.

These are the most remote from each **other**: for, as just observed, they include as many contradictions as there are individuals denoted by the subject. **It** is obvious that they cannot **both be** true. **But** they may both be false; for each asserts more than the mere negation of the other, and this superfluous assertion may be false.

127. Thus it may **be** false that 'All men are black,' and therefore true that 'Some men **are** not **black.**' But the contrary asserts more than this, and the additional assertion is false.

128. Hence contraries cannot **both be true**, but both may be false, **and the** rule of inference is:

From the **truth of one** contrary we **may infer the falsity** of the other;

but,

From the falsity of one **contrary** we cannot infer the truth **of** the other.

129. These rules may be deduced from those of subalternation and contradiction:

If A is true, I is true; therefore E false.
If A is false, O is true; thence no inference as to E.
If E is true, O is true; therefore A false.
If E is false, I is true; thence no inference as to A.

130. Subcontrary propositions are those **of which one affirms** particularly what the other denies particularly. They must be I and O:

Some men are black;
Some men are not black,

are subcontraries.

131. Both may be true; both cannot be false. For if it is false that, 'Some men are black,' its contradictory, 'No men are black,' is true; and therefore its subalterna, 'Some men are not black,' is true. In general, if I is false, its contradictory E is true, and therefore O is true. If I is true, E is false, and no inference can be drawn as to O.

132. Hence the rule of inference is:

From the falsity of one subcontrary we may infer the truth of the other;

but,

From the truth of one subcontrary we can draw no inference as to the other.

133. Subcontrariety is not properly called opposition, since the subjects of the subcontraries are not the same; the 'some' of one not being identified with the 'some' of the other. It is therefore called apparent opposition.

134. The relations of subalternation and opposition may be exhibited in a diagram:

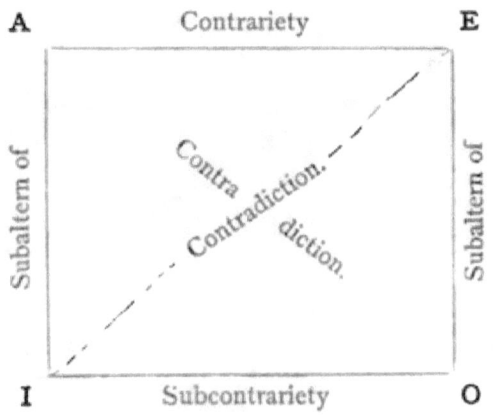

Of Conversion.

135. Conversion is **the transposition of the subject and predicate, so that the subject becomes predicate and the predicate becomes subject.**

The original is called the **Convertend.**
The derived is called the **Converse.**

136. Conversion **is Logical when** the Converse follows from the Convertend.

137. Conversion **is of three kinds**; simple, per accidens, and by Contraposition.

138. Conversion **is simple** when the quantity of the Converse is the same as the quantity of the Convertend, **as:**

> Some **men are** black (I);
> Some **black** things **are** men (I).

139. Conversion is per **accidens** when the Convertend is universal and the Converse is particular, as:

> All men are **mortal (A)**;
> Some mortal things **are** men (I).

140. I is converted simply (into I):

From the proposition, 'Some lawyers are historians,' we may infer, 'Some historians are lawyers.' 'Some As are B' is converted to 'Some Bs are A.' Both statements are equivalent to, 'Some things are both A and B (or are AB').

141. E is also converted simply (into E):

From the proposition, 'No savages are philosophers,' we may infer, 'No philosophers are savages.' 'No A is B' is converted to 'No B is A.' Both statements express the same thing, viz.: 'Nothing is at once A and B (or is AB'), or the extension of A and that of B are mutually exclusive.

142. A cannot be converted simply, but is converted per accidens into I:

From the proposition, 'Every man is mortal,' we cannot infer, 'Every mortal thing is a man.'

143. It has already been shown that the predicate of an affirmative proposition is particular; thus, in the proposition, 'All men are mortal,' 'mortal' is taken in part of its extension; 'Men are amongst mortal things.' Hence when 'mortal' becomes the subject it is still particular, otherwise we should be reasoning from the particular to the universal, and therefore the proposition of which it is the subject is particular, 'Some mortal things are men.'

144. If we regard the mark of quantity, 'every,' 'all,' 'some,' as belonging to the subject, as it really does, we shall see that it is the subject of the original proposition that in becoming predicate loses its universality, and that, because we do not express the quantity of the predicate.

145. If we adopt the view that the predicate of an affirmative proposition may be quantified, we shall be able to convert **A simply**. 'All men **are mortal**' becomes 'Some mortal things **are all men,' or, in** common language, 'Only mortal things **are** men.'*

146. O cannot **be** converted while **it retains its** quality, **for** its subject is particular ; and if made the predicate of the converse (which would **be** negative), it must needs be universal. **We** should thus be reasoning from part **of the** extension of a term **to** the whole (*a particulari ad universale*), which **is** invalid. **Thus**, from 'Some men are not learned' **we cannot infer 'Some** learned beings are not men,' nor any other proposition **of which** 'some learned' is the subject. For nothing whatever has been asserted about learned beings.†

147. In order to convert **O**, we must therefore treat it as I, which we have seen we can **always** do by attaching the negative particle to the predicate (62). We then convert it **by** Contraposition.

Of Contraposition.

148. **Contraposition or Conversion by Contraposition consists in substituting for predicate and copula their contradictories and then converting.**

---

* It would usefully simplify the doctrine of conversion and **connected** processes if quantification **of the** predicate were admitted **in** this **case** at least, in which the Convertend logically leads to a universal predicate in the Converse.

† Logicians who adopt the '**quantification of** the predicate' convert O thus : "Some As are not **B**' is converted to ' No B is some A'; *ex. gr.* '**Some men** are not learned' **is** converted to 'No learned **beings** are **some men.**'

149. A may be contraposed simply, for its equipollent is E.

> Every A is B

is equipollent with

> No A is not B;

and this is converted to

> No not-B is A = Whatever is not B is not A.

150 O may also be contraposed simply, since it is equipollent with I.

> Some As are not B

is converted into

> Some things not B are A.

Thus:

> Some men of genius are not learned

becomes

> Some unlearned men are men of genius.

151. E by equipollence is A, and therefore is only contraposed per accidens.

> No A is B

becomes

> Some things not B are A.

Thus:

> No idler is successful

becomes

> Some of the unsuccessful are idlers.

152. I by equipollence becomes O, and therefore is not contraposed.

Remarks on Conversion.

153. (*a*) A universal affirmation sometimes seems **as** if it **were** convertible simply **(A to A)**. All equilateral triangles **are** equiangular, and, All equiangular triangles are equilateral. But the latter proposition **is not** inferred from the form of the first, but from the matter, *i. e.* from something we know, but which is not expressed.

154. There are, however, some cases, **as shown above** (86), in which **the** predicate is expressly made universal by the words 'only,' 'alone,' etc. ; also where the proposition takes one of the forms that express identity (89), **as:**

A is equal to B.

155. (*b*) **In** the case **of** propositions indicating dependence of **one** attribute on another **(93), the** meaning of the Convertend and that of the Converse are not identical; *ex. gr.* 'Virtue is happiness' **is not the** same as 'Happiness is virtue'; 'Whatever is right is alone expedient' (or, 'All the right is all the expedient') is different from 'Whatever **is** expedient **is** alone right; **(or,** ' All the expedient is all the right ') ; **or,** 'Whatever **is** not expedient **is** not right.' All these propositions affirm that right and expediency go together; but **the** question is, by which **quality** are **we** to determine the presence of the other. **One** proposition states that expediency is the test of **right, the** other (the first) states that right is the test of expediency. No error arises from this difference, because it is usually clear which of the two attributes is supposed to **be most** easily recognized.

156. (*c*) 'Converse' and 'conversely' **are** often used **with** reference **to** two propositions in **which,** instead of the Copula we have a word (or words) expressing a more complex relation, and this word remains while the related terms change places; *ex. gr.* 'The truth of either contradictory follows from the falsity of the other, and conversely, the falsity of either follows from the truth of the other.' The student must guard against confounding this with logical conversion. It would be more correct in such cases to use the term '*vice versâ.*'

Of the Principle of Substitution.

157. There are some immediate inferences which, although obviously valid, and of frequent occurrence, present some difficulty when we try to bring them under any of the preceding heads. For example: 'A negro is a man; therefore, a sick negro is a sick man.' Or, again: 'A horse is a sensitive animal; therefore, he who tortures a horse tortures a sensitive animal.'

158. The principle which is employed in these cases is an extension of the principle of Subalternation. It is this:

> **For any term used universally (*i.e.* distributively) less may substituted.**

Thus, wherever the name of the genus is used universally the name of the species may be substituted. This is obviously implied by the distribution of the term. Whatever is true of 'every' is true of each.

159. Accordingly, in the identical proposition, 'A sick man is a sick man,' we may substitute for 'man' in the subject, 'negro,' which is part of its extension. Similarly, in the identical proposition, 'He who tortures a sensitive

animal tortures a sensitive **animal**,' we substitute for ' sensitive animal' in the subject, where it is distributed, the term ' horse,' which is part of its extension.

160. On the other hand, it is a principle that—

'**For any** term used **particularly we may substitute a term of wider** extension.'

For it is clear that whatever is part of the less extension is part of the greater,

161. In both cases of substitution it **is of** course assumed that **the words** employed have a definite signification, which **is the** same whatever **the** connexion in which they occur. This is not the case with words **implying** comparison. Thus: from 'a cottage is a house' we cannot infer 'a **huge** cottage is a huge house '; **or**, again : from 'a tailor is a man' we cannot infer that 'the best of tailors is the best of **men**.'

CHAPTER II.

Of Mediate Reasoning, or Syllogism.

162. Mediate Reasoning is the deduction of one proposition from two **or** more. All mediate reasoning may be reduced **to Syllogism,** which is the inference of one **proposition from two, in** which, taken together, **it is** by the mere form of expression necessarily involved. *Ex. gr.:*

   Logic is a Science ;
   All Sciences **are** worthy of study ;
therefore,
   Logic **is** worthy of study.

Here the reasoning does not depend on the meaning of the words, but is equally valid if we substitute symbols:

>S is M,
>Every M is P;
>
>therefore, S is P.

163. The conclusion before it is proved is called the **Question**. The propositions from which it is deduced are the **Premisses** (*propositiones praemissae*). The **Subject** of the Conclusion is called the **Minor Term**. Its predicate is the **Major Term**. Both are called **Extremes**.

164. Syllogisms are either simple or complex. Simple syllogisms (otherwise called categorical) consist of simple propositions, *i. e.* such as assert absolutely, that is, without a condition.

165. In simple syllogisms, the coincidence or non-coincidence of the extremes is ascertained by comparing them with a third term, which is called the **Middle Term**.

166. The premiss in which the major term occurs is called the **Major Premiss**; that in which the minor term occurs is the **Minor Premiss**. The order of the premisses is of no consequence; but logicians commonly state the major first.

167. In representing syllogisms symbolically it is usual to employ S. (the initial of Subject) for the minor term, which is Subject of the conclusion; P for the major term, which is Predicate of the conclusion, and M for the Middle. The conclusion, therefore, always appears as SP. The symbol ∴ is used for 'therefore' (∵ is sometimes used for 'because').

General Rules of Syllogism.

168. The general rules are those which are applicable to all syllogisms :—

(1.) The middle term must be at least once universal; in other words, it must be distributed.

For if it were undistributed (*i. e.* not taken universally), it might be taken in two different parts of its extension, and there would be really two middle terms. Thus in : All men are animals; All horses are animals : the parts of the extension of 'animals' are different, and therefore we can deduce nothing as to the relation between 'men' and 'horses.'

If this rule is violated, we have the fallacy of '**Undistributed Middle.**'

169. (2) An extreme must not be taken more universally in the conclusion than in the premisses; in other words, if a term is universal in the conclusion, it must have been universal in the premisses.

For otherwise we should be arguing from the part to the whole, or as it is called, *a particulari ad universale.*

The violation of this rule is called **Illicit Process**, which may be either of the major or the minor.

170. *Cor.* Hence and from the preceding rule it appears that there must be at least one more universal term in the premisses than in the conclusion. For any term which is universal in the conclusion must have been universal in the premisses, and the middle term, in addition, must be once universal.

171. (3) From two negative premisses nothing follows. For if nothing is asserted in the premisses except that the extremes are both excluded from the middle term, nothing has been implied as to their relation to each other.

172. (4) From two affirmative premisses a negative conclusion cannot follow.

For if what we assert of each of the extremes is either that it contains or is contained by the same middle (and therefore partially coincides with it), it is not thereby implied that either extreme excludes the other.

173. (5) If either premiss is negative, the conclusion is negative.

For when what is asserted is only that one extreme either contains or is contained by the middle, which is excluded from the other extreme, it is not thereby implied that either extreme contains the other; hence an affirmative conclusion cannot follow.

174. (6) From two particulars nothing follows.

They must be either II or OI or IO. In II all the terms are particular; therefore the middle is undistributed, contrary to the first rule.

In OI or IO there is only one universal term, viz. the predicate of O, which must be the middle term; therefore there is no universal term in the conclusion. But as one premiss is negative, the conclusion, if any, would be negative, and therefore the major term in it universal. The major term, therefore would be particular in the premiss and universal in the conclusion, contrary to the second rule.

175. (7) If either premiss is particular, the conclusion is particular.

First Case: If both premisses are affirmative, they are A and I. Here there is only one universal term, viz. the subject of **A**. **This must be the** middle term; therefore there can **be no universal term in** the conclusion, which must be **I**.

Second Case: One premiss is negative. **In** this case **there are** only two universal terms, viz. the predicate of **the** negative and **the** subject of **the** universal; therefore there **is** but one in the conclusion (*Cor.*). But **as** the conclusion must be negative, its predicate will **be** universal **and its subject** particular, and **this** makes the conclusion particular.

176. This rule and **the** fifth are sometimes expressed together, thus: The conclusion follows **the weaker** part, the negative being considered weaker than the affirmative, and the particular weaker than the universal.

177. Although we **cannot** draw any conclusion in extension **from** premisses **with** undistributed middle, yet we may sometimes **draw** an inference **in** Attribution. An example will **show** what this means: '**Men** have eyes; insects have eyes; therefore **men** have some attributes in common with **insects**.' Here **we reason** not about the classes men and **insects** having or not **having** individuals in common, but about the groups of attributes. The argument may be brought into the usual form by making the attribute the subject of both propositions, thus: To have eyes is an attri**bute** of men; to have eyes is an attribute **of** insects; therefore some attribute of men **is** an attribute **of** insects.

Of the **Four** *Figures.*

178. **The Figure of Syllogism is the disposition of** the **middle** term in the **premisses.**

179. **In** the first figure the middle **is** subject of the major premiss, **and** predicate of the minor. **In** this **case the** ex-

tremes have the same position in the conclusion as in the premisses,

$$S \text{ is } M;$$
$$M \text{ is } P;$$
$$\therefore S \text{ is } P.$$

Ex. gr. Logic is a Science;
All Sciences are worthy of study;
∴ Logic is worthy of study.

180. The principle of this figure is: Whatever is universally predicated (affirmatively or negatively) of a common term may be similarly (*i.e.* affirmatively or negatively) predicated of anything contained under that term. It may be expressed otherwise, thus: Whatever comes under the condition of a rule comes under the rule.

$$\text{Every } M \text{ is } P \text{ (or not } P);$$
$$S \text{ is } M$$
$$\therefore S \text{ is } P \text{ (or not } P).$$

181. In the second figure the middle is predicate of both premisses. In this case the major term is differently placed in the premiss and in the conclusion:

No fish breathe in air;
Whales breathe in air;
∴ Whales are not fish.

Or,

All fish breathe in water;
Whales do not breathe in water;
∴ Whales are not fish.

182. The principle of this figure is: If an attribute is

predicated affirmatively or negatively of every member of a class, then any subject of which it cannot be so predicated does not belong to that class:

> Every P is M (or not M);
> S is not M (or is M);
> ∴ S is not P.

183. In the third figure, the middle term is subject of both premisses. In this case the minor term occupies a different place in the premiss and in the conclusion; *ex. gr.*:

> Bats fly;
> Bats are not birds;
> ∴ Some things that fly are not birds.

184. The principle of this figure is: If anything which belongs to a certain class possesses a certain attribute (positive or negative), then that attribute is not incompatible with the attributes of that class:

> M is S;
> M is P (or not P)
> ∴ Some S is P (or not P).

(This represents M as singular; if it is a common term, of course, in one premiss we must have 'Every M.')

185. In the fourth figure the middle term is predicate of the major premiss, and subject of the minor. In this case both extremes are differently placed in the premisses and in the conclusions; *ex. gr.*:

> Every P is M;
> No M is S;
> ∴ Some S is not P.

186. The diagrams of the four figures are:

First.	Second.	Third.	Fourth.
MP	PM	MP	PM
SM	SM	MS	MS
SP	SP	SP	SP

Special Rules of the Four Figures.

187. Special Rules are those which are applicable only to particular figures.

188. The Special Rules of the **First Figure** are:

(1) The minor must be affirmative.
(2) **The major must** be universal.

189. (1) The minor must be affirmative. For if it were negative, the major must be affirmative and its predicate particular; and the conclusion would be negative and its predicate universal; but the predicate of the major premiss and the predicate of the conclusion are the same, viz. the major term, which would thus be particular in the premiss and universal in the conclusion, contrary to the second general rule.

190. (2) The major must be universal. For, since the minor is affirmative, its predicate, which is the middle term, is particular; it must, therefore, be universal in the major premiss, where it is the subject, and makes that proposition universal.

191. NOTE.—These rules are manifest at once from the diagram:

MP
SM
―
SP

If P is universal in the conclusion, it must have been universal in the premiss, *i.e.* if the conclusion is negative, it was the major premiss that was negative. The second rule is seen at once from the position of M. In this figure the major is the general principle, and the minor brings a case under the condition of the rule.

192. The Special Rules of the Second Figure are:

(1) **One of the premisses (and therefore the conclusion) must be** negative.

(2) **The** major premiss must be universal.

193. (1) One of the premisses must be negative. For the middle term is predicate of both; and if both were affirmative, the middle would be undistributed.

194. (2) The major must be universal. For, since one conclusion is negative its predicate is universal; therefore, it must have been universal in the major premiss, where it is subject, and makes that premiss universal.

195. NOTE.—These rules are manifest from the diagram:

$$\frac{\begin{matrix}PM\\SM\end{matrix}}{SP}$$

196. The Special Rules of the **Third Figure** are:

(1) **The minor must** be affirmative.

(2) **The conclusion must be** particular.

197. (1) The minor must be affirmative. This is proved as in the first figure, the position of the major term being the same in these two figures.

198. (2) The conclusion is particular. For the minor

being affirmative, its predicate, which is the minor-term, is particular; it is therefore particular in the conclusion, of which it is subject, and therefore the conclusion is particular.

NOTE.—These rules will be seen at once from the diagram:

$$\frac{\begin{array}{c} MP \\ MS \end{array}}{SP}$$

200. The Special Rules of the **Fourth Figure** are:
(1) **When the major is affirmative, the minor is universal.**
(2) **When the minor is affirmative, the conclusion is particular.**
(3) **In negative moods** (*i.e.* when the conclusion is negative), **the major is universal.**

201. (1) When the major is affirmative the minor is universal. For if the major is affirmative, its predicate, which is the middle term, is particular; therefore it must be universal in the minor premiss, where it is subject, and makes that premiss universal.

202. (2) When the minor is affirmative the conclusion is particular. For the predicate of the minor premiss is subject of the conclusion, and if particular in the premiss (*i.e.* if the minor is affirmative), it is particular in the conclusion (*i.e.* the conclusion is particular).

203. (3) In negative moods the major must be universal. For if the conclusion is negative, its predicate, the major term, is universal; therefore it must have been universal in the premiss, where it is subject, and therefore that premiss is universal.

204. The diagram makes these rules manifest:

$$\frac{\begin{array}{c}\text{MP}\\ \text{MS}\end{array}}{\text{SP}}$$

NOTE.—These rules are all hypothetical; because each term being once subject and once predicate, we are compelled to argue from quality to quantity, or *vice versâ*. Of course the rules might be converted, and we might say, *ex. gr.* If the minor is particular the major must be negative, &c.

Remark.

205. It will be observed that the first figure is the only one in which the conclusion A can be drawn. In the second figure the conclusion must be negative; in the third, particular, in the fourth, either negative or particular. We may prove this directly as follows:

206. If the conclusion is A the premisses must be A A (by the general rules). The only universal terms here are the two subjects. Now as the minor is universal in the conclusion A, it must have been universal in the minor premiss; therefore it is its subject, and the predicate which is particular is the middle term; this term must therefore be universal in the major premiss, and is therefore its subject. The syllogism then is in the first figure.

207. This is not of itself a reason for regarding the first figure as preferable to the others. If our premisses justify only a negative or particular conclusion, then to prefer the first figure, because with other premisses it would give a conclusion A, is like choosing to walk on the high road when it is not the nearest way, because it is wide enough for a coach and four.

208. On the other hand, if we have a premiss O we cannot use the first figure (except by taking instead of O its equipollent I).

209. In fact each of the three first figures has its appropriate use. The first must be used when we have to establish an A conclusion, and in general when we refer a case or class of cases to a general rule.

210. The second figure is used when we wish to prove that certain things do not belong to a certain class, because they differ from that class either by the possession or the want of some attribute. The name of this attribute is naturally predicated of both extremes.

211. The third figure is useful when we wish to disprove a proposition alleged or assumed to be universal. This we do by establishing an exception to it. In such arguments the middle is likely to be either a singular term or the name of a definite class, and as a singular or definite term can only be subject of a proposition, the argument falls naturally into the third figure.

212. The fourth figure alone has no appropriate use, being in fact a perverse mode of expressing reasoning which falls naturally into another figure.

Of Moods.

213. **Mood is the determination of the propositions of the syllogism as to quantity and quality.**

214. As each of the three propositions, if considered by itself, may have four varieties, A, E, I, O, the number of arrangements arithmetically possible is $4 \times 4 \times 4 = 64$. But most of these are logically impossible as violating the general rules. The number of legitimate moods may be ascertained logically as follows:

215. If the major is A the minor may be A, E, I, O. If the major is E the minor must be A or I. If the major is I the minor must be A or E. If the major is O the minor must be A. Hence we have nine possible sets of premisses. Now as to the conclusion: if the premisses are universal (AA, AE, EA), the conclusion may be either

particular or **universal**; in all other cases the quantity and quality of the conclusion are determined by those of the premisses. We have, therefore, twelve legitimate moods; but only eleven of these are useful, for the mood **AEO** is always useless (when legitimate), since as the **minor** term is universal in the minor premiss, whether it be subject or predicate, the conclusion may be E. The mood **EAO** is useless in the first and second figures, as the **minor term is** subject of its premiss, and therefore the conclusion may be E. Lastly AAI is useless in the first figure, where the premisses AA justify a conclusion **A**.

216. Now, let us test these moods by the special rules, in order to see which of them are admissible in each figure. The first figure can have as major only A or E, and as minor **A or I**.

Hence we have the four moods,

AAA, AAI, EAE, EIO.

217. The second figure may have A or E as major. If major be **A** the minor is E or O, and the conclusion E or O accordingly. If major be **E**, minor may be A with conclusion E, or I with conclusion O.

Hence we have as moods of the second figure

AEE, AOO, EAE, EIO.

218. In the third figure the minor is A or I. The major is unrestricted, but the **conclusion** is particular.

Hence the moods are

AAI, AII, EAO, EIO, IAI, OAO.

219. In the fourth figure the major may be A, E, or I. With major A the minor is A or E (first special rule). With minor A conclusion is I (second special rule); and with minor E conclusion is E (O being useless).

PART III.]   *OF INFERENCES.*   53

With major E minor is A or I; and in either case the conclusion is O (second rule). With major I minor is A (third rule) and conclusion I.

Hence we have in the fourth figure

AAI, AEE, EAO, EIO, IAI.

220. These moods all have settled names which are contained in the following mnemonic lines:

*Barbara, Celarent, Darii, Ferio*que prioris
Cesare, Camestres, Festino, Baroko secundæ:
Tertia *Darapti, Disamis, Datisi, Felapton,
Bokardo, Ferison,* habet; quarta insuper addit
Bramantip, Camenes, Dimaris, Fesapo, Fresison.

The vowels of these names indicate the nature of the propositions; some of the consonants have a signification which will appear presently.

Remark.

221. Strictly speaking we have only proved that these moods do not contradict any of the rules. But that they are also all conclusive in their proper figures appears at once from a comparison with the rules of the figures.

222. In the first figure something is in the major predicated universally of a class (A or E); in the minor something is asserted to belong to that class (A or I), and hence in the conclusion we may predicate of it what we have predicated of the class, the quality of the conclusion being the same as of the major premiss, its quantity that of the minor.

223. In the second figure we assert in the minor that something possesses universally or particularly (A or I) an attribute which a certain class lacks (E), or else that it lacks (E or O) an attribute which the class possesses (A). Hence we may conclude that the thing spoken of is excluded from that class, universally or particularly, according to the quantity of the minor.

224. In the third figure we assert in the major that certain things possess a certain attribute, positive or negative (A, E, I, or O), and in the minor that they belong (A or I) to a certain class. (The distribution of the middle insures that we speak of the same things.) Hence we may infer that some members of this class possess the given attribute, positive or negative (I or O).

225. In the fourth figure the moods Bramantip, Camenes, Dimaris, are the same respectively as Barbara, Celarent, and Darii of the first figure, only that the conclusion, instead of being the direct conclusion of the first figure, is its converse. The remaining moods of the fourth figure, Fesapo and Fresison, are really the same as Felapton and Ferison, respectively, in the third figure, only that instead of the real major its simple converse is used. These five moods then are valid.

Examples of the Moods.

226. Examples of these moods follow:

227. First Figure.

Every M is P	Bar		Every M is P	Da
Every S is M	ba		Some S is M	ri
∴ Every S is P	ra		∴ Some S is P	i

No M is P	Ce		No M is P	Fe
Every S is M	la		Some S is M	ri
∴ No S is P	rent		∴ Some S is not P	o

228. Second Figure.

No P is M	Ce		Every P is M	Cam
Every S is M	sa		No S is M	es
∴ No S is P	re		∴ No S is P	tres

No P is M	Fes		Every P is M	Ba
Some S is M	ti		Some S is not M	ro
∴ Some S is not P	no		∴ Some S is not P	ko

OF INFERENCES.

229. Third Figure.

Every M is P	Da	Every M is P	Da
Every M is S	rap	Some M is S	tis
∴ Some S is P	ti	∴ Some S is P	i

No M is P	Fe	No M is P	Fe
Every M is S	lap	Some M is S	ris
∴ Some S is not P	ton	∴ Some S is not P	on

Some M is P	Dis	Some M is not P	Bok
Every M is S	am	Every M is S	ar
∴ Some S is P	is	∴ Some S is not P	do

230. Fourth Figure.

Every P is M	Bram	Some P is M	Dim
Every M is S	an	Every M is S	ar
∴ Some S is P	tip	∴ Some S is P	is

Every P is M	Cam
No S is M	e
∴ No S is P	nes

No P is M	Fes	No P is M	Fres
Every M is S	ap	Some M is S	is
∴ Some S is not P	o	∴ Some S is not P	on

Of Aristotle's Dictum.

231. Aristotle (and after him other logicians) sought to bring all mediate reasoning under a single principle, to show, in fact, that the process is in every case essentially one and the same.

232. According to this view all reasoning consists in applying a general rule to a particular case or group of cases. The general rule is the universal major; the

affirmative minor brings the particular case under the condition of the rule, and in the conclusion it is inferred that it comes under the rule itself. This, as will be seen presently, is not limited to simple syllogisms.

233. As applied to these, the formal principle of the reasoning may be stated thus:

Whatever belongs to an attribute of a thing belongs to the thing itself.

Or as it is stated by Aristotle—

Whatever is said of the **predicate shall be said also of the** subject.*

234. From this can be deduced the ordinary form of the Dictum de Omni et de Nullo, ' Whatever belongs to or contradicts the class belongs to or contradicts all the objects contained under the class.' For the meaning of the class name is the group of attributes belonging to the class.†

235. This principle is directly applicable to the first figure only; and in order to show that it applies to the other figures we must reduce them to the first figure.

236. The problem of reduction in any given case is: given premisses and a conclusion, which, as they stand, are not in the first figure, to deduce from the given premisses the required conclusion by a syllogism in the first figure. The mood to be reduced is the reducend; that to which it is reduced is the reduct.

---

* "Ὅσα κατὰ τοῦ κατηγορουμένου λέγεται πάντα καὶ κατὰ τοῦ ὑποκειμένου ῥηθήσεται.

† The Dictum is given by Aristotle as a definition of κατὰ παντὸς κατηγορεῖσθαι.

Of Reduction.

237. Reduction is ostensive when the conclusion obtained in the first figure is either the same as that required or yields it by conversion.

238. Any mood which has not O as a premiss may be readily brought into the first figure by Ostensive **reduction**. If O is a premiss it must be treated as I.

239. In the second figure (PM, SM, SP), the premisses do not of themselves show which term is predicate and which subject of the conclusion. If the given minor is affirmative, we have only to convert the major simply (it being always universal and in this case negative). This applies to Cesare and Festino; the *s* indicates simple conversion of the preceding E, and the initials show that the moods are reduced to Celarent and Ferio respectively. Thus:

No P is M	Ex. gr. *No fish breathe in air:*
∴ No M is P	∴ Nothing breathing in air is a fish:
S (some or all) is M	Whales breathe in air:
∴ S (some or all) is not P.	∴ Whales are not fish.

(N. B.—In this and other examples the propositions which do not belong to the first figure are in italics.)

240. If the given minor is E it cannot be a minor in the first figure; it must, therefore, be treated as major and converted simply. The conclusion, which will be E, will have to be converted, in order to give the required conclusion. This applies to Camestres. In this name, *m* indicates the interchange of premisses ('metathesis'); *s* in the premisses the conversion (simple) of the preceding E; *s* after the conclusion, the conversion not of this conclusion but of that obtained in the first figure. The **reduct mood** is Celarent. Thus:

 Every P is M   *Ex. gr.* All fish breathe in **water**:
 No *S is M*   No whales breathe **in water** :
 ∴ No M is S   ∴ Nothing breathing in water is a **whale**:
 ∴ No P is S   ∴ No fish is a whale:
 ∴ *No S is P.*   ∴ *No whale is a fish.*

241. This might be reduced **by** converting the major by contraposition. Thus:

 Every P is M   *All fish breathe in water* :
 ∴ No not-M is P   ∴ Nothing that does not breathe in
             water is a **fish** :
 Every S is not-M   Whales do not breathe in water:
 ∴ No S is P.   ∴ Whales are not fish.

242. If the given minor is O, it cannot stand **in the first** figure either as major **or** minor. We must, therefore, treat **it as** I, **and convert the** major **by** contraposition. Thus:

  All poets are men of genius :
 ∴ **No** man not a man of genius is a poet :
   Some rhymesters are not men of genius :
 ∴ **Some** rhymesters are not poets.

    All P is M
   ∴ No not-M is P
    Some S is not M
   ∴ Some S is not P.

The reduct is in Ferio; and if *c* were used to indicate conversion by contraposition, and *y* to indicate the change of O to its equipolent I, the name might be Facoyro. This corresponds to the second mode of reducing Camestres.

243. In the third figure (MP, MS, SP), as in the second, the premisses do not determine which is the subject and which the predicate of the conclusion. If the major is universal, we have only to convert the minor, which is always affirmative. This is the case in Darapti, Datisi Felapton, and Ferison. The *s* in Datisi and Ferison indicates simple conversion of the I minor; *p* in Darapti and Felapton indicates conversion of the A minor per accidens.

244. If the major is particular, it cannot stand as such in the first figure; it must be treated as a minor, and, as in the preceding case, converted; if I, simply; if O, by contraposition; and the conclusion will give by conversion the conclusion of the proposed syllogism. This is the case in Disamis and Bokardo. In order that the name of the latter should indicate this mode of reduction it should be Docamos, using *c*, as before, to indicate conversion by contraposition. In this case the conclusion obtained in the first figure is I, and has to be simply converted; and the negative, which will then be in the predicate, must be attached to the copula, but this involves no change in expression. Thus:

Every M is P (or not P)	All ants are invertebrate (or not vertebrate):
M (all or some) is S	*Ants (all or some) are sagacious:*
∴ Some S is M	∴ Some sagacious things are ants:
∴ Some S is P (or not P).	∴ Some sagacious things are invertebrate (or not vertebrate).

245. This example represents Darapti, Datisi, Felapton, and Ferison. If the conclusion were, 'Some invertebrates are sagacious,' it would be Disamis.

The following represents Disamis and Bokardo:

Some M is P	*Some histories are amusing:*
∴ Some P is M	∴ Some amusing books are histories:
Every M is S	All histories are instructive;
∴ Some P is S	∴ Some amusing books are instructive:
∴ *Some S is P.*	∴ *Some instructive books are amusing.*

By reading 'not P' for 'P,' we have Bokardo:

Some M is not P	Some histories are not amusing:
∴ Some not-P is M	∴ Some books not amusing are histories:
Every M is S	All histories are instructive:
∴ Some not-P is S	∴ Some books not amusing are instructive:
∴ *Some S is not P.*	∴ *Some instructive books are not amusing.*

The process is obviously the same in substance in all the moods.

246. When the middle term is singular, or the name of a definite group, the reduction of the third figure is awkward and unnatural; *ex. gr.*:

> X is a recent poet:
> X is a poet of the first rank;
> ∴ Some recent poet is of the first rank.

In order to reduce this we must convert the minor into,

> One recent poet is X.

The awkwardness appears in the first of the previous examples. We really predicate of ants that they are sagacious, not of sagacious things that they are ants.

247. In the fourth figure (PM, MS, SP), if the major is affirmative and the minor consequently universal, they must be treated as minor and major respectively, *i. e.* we make the conclusion PS, which may be then converted into SP (for it will not be O; since in negative moods the given

major is universal, and therefore P may be universal in the conclusion). This is the case in Bramantip, Camenes, and Dimaris. The *p* in Bramantip means that the conclusion A obtained in Barbara (to which this mood is reduced) is converted per accidens.

248. If the major is negative (E) it cannot be minor in the first figure; in this case, therefore, both premisses are converted. This is the case of Fesapo and Fresison.

249. It will have been observed that the initial letter of the mood of the reducend is the same as that of the reduct; that *s* shows that the premiss preceding is to be converted simply; *p* that it is to be converted per accidens; *m* that the premisses have to be transposed, or rather that they change names, since the order is indifferent. In the conclusion *s* and *p* refer to the conclusion in the first figure, which has to be converted to give the required conclusion.

Of Reductio ad Impossibile.

250. The names of Baroko and Bokardo indicate a different and more complex mode of reduction, viz.: the contradictory of the conclusion is substituted for the O premiss, and from this and the retained premiss A we deduce in the first figure and mode Barbara a conclusion, A, which contradicts the given O premiss. Since this conclusion contradicts a premiss which is given true, it must be false, therefore one of the premisses from which we inferred it is false; and, as one was given true, the falsity must be in the substituted premiss, which contradicts the required conclusion. Therefore this conclusion is true. This reduction is called **Reductio ad** impossibile.

251. Stated more shortly the reasoning would stand thus:

Baroko: Every P is M:
Some S is not M:
∴ Some S is not P.

This is reduced to,
Therefore, if
then (by Barbara),

> Every P is M.
> Every S is P,
> Every S is M.

But this is false, for Some S is not M; ∴ it is false that Every S is P, *i.e.* it is true that Some S is not P.

252. This is unsatisfactory, for we have here got a conditional syllogism, the reduction of which to a categorical form would give us back the original syllogism.

Of the Unfigured Syllogism.

253. Those logicians who adopt the principle of **quantifying the predicate** escape the complexity of the figures. For, as already shown, every proposition is on that view an equation or a statement of inequality. These logicians consequently found **syllogism on** the following axioms:—

Terms which coincide as to their extension with a third term coincide with each other.

Terms of which one coincides with, and the other is excluded from, the extension of the same third term are excluded from each other.

If two terms are both excluded from one and the same **third term, we can infer nothing as to their** relation to **each other.**

254. On this doctrine all categorical syllogisms are of the type,

> A is equal to B:
> B is equal to C:
> ∴ A is equal to C.

Or for negatives—

> A is equal to B.
> B is not equal to C:
> ∴ A is not equal to C.

The order of the terms in each premiss is indifferent, and consequently there is no distinction between major and minor.

255. It should be observed that whatever opinion is adopted about the quantification of the predicate in general, this is the form of the syllogism in mathematical reasoning, and in other cases where the premisses are propositions in identity.

Of the Enthymeme.

256. An **Enthymeme** is a syllogism of which one premiss is suppressed, as : A is B ∴ A is C. The expressed premiss is called the **Antecedent**, and the conclusion is called the **Consequent**. If the subject of the consequent appears in the antecedent, this is the minor, and the major is suppressed ; if the predicate of the consequent appears in the antecedent, this is the major, and the minor is suppressed.

NOTE.—This is the commonly accepted meaning of 'enthymeme,' but it seems to have originated in a false etymology, as if the word were derived from ἐν θυμῷ, from one premiss being in the mind. It is really derived from the verb ἐνθυμέω. Aristotle used it to mean an argument from signs and likelihoods.

Of Sorites.

257. A **Sorites** is a chain of reasoning consisting of a series of syllogisms in which each intermediate conclusion is not expressed, but is assumed as a premiss of the succeeding syllogism.

Usually the syllogisms are in the first figure, and then the predicate of each premiss is the subject of the next, and the predicate of the last is in the conclusion predicated of the first subject. *Ex. gr.:*

(Suppressed conclusions used as minor premisses.)

 A is B
 Every B is C ∴ A is C :
 Every C is D ∴ A is D.
 Every D is E
∴ A is E.

258. **The Sorites** in the first figure has two **Special Rules.**

> **The first premiss alone can be particular.**
> **The last premiss alone can be negative.**

259. The first alone can be particular. When the Sorites is resolved into syllogisms, the first premiss is the minor of the first syllogism, but the minor of every other is the (suppressed) conclusion of the preceding; for the subject of the first premiss is the subject of every conclusion. All the expressed premisses after the first, being therefore majors in the first figure, must be universal.

260. The last alone can be negative. For if any premiss were negative, the conclusion of the syllogism would be negative; but it is the minor of the succeeding one, and therefore must be affirmative.

CHAPTER III.

Of Complex Propositions.

261. Complex Propositions are those which combine two or more simple propositions in such a way that the truth or falsity of one is said to depend on the truth or falsity of the other or others. They are divided into Conditional (or Hypothetical) and Disjunctive.

262. A **Conditional** Proposition is one which asserts that the truth of one proposition depends on the truth of another.

>If A is B, it is C :
>If A is B, C is D :
>If A is B, either C is D or E is F.

The dependent proposition is called the **Consequent**, that on which it depends is the **Antecedent**.

263. The truth of a Conditional proposition does not depend on the truth of the separate propositions; it only requires that the consequent follow from the antecedent.

264. The antecedent may be related to the consequent, as Reason to Consequence or as Cause to Effect. *Ex. gr.*:

>If there is dust in the air, a beam of light passing through it is visible (Relation of Cause to Effect).
>If a beam of light is visible, there is dust in the air (Relation of Reason and Consequence).

265. A conditional which has only three terms may often (if these are common terms) be reduced to a simple pro-

position. 'If A is B it is C' cannot express a consequence formally conclusive unless 'Every AB is AC'. Thus, the first of the preceding propositions is equivalent to, 'Dusty air makes light visible'; and the second to, 'Dusty air alone makes light visible'.

The converse by contraposition of 'Every AB is AC' is 'Whatever is not AC is not AB', equivalent to, 'If A is not C it is not B'. This may, therefore, be called the converse of the original conditional.

266. In a **Disjunctive Proposition** the truth of one of the component propositions depends on the falsity of the other (or others); *ex. gr.*:

> A is either B, or C, or D :
> Either A is B, or C is D, or E is F.

267. Disjunctives may be reduced to conditionals, thus: 'A is either B or C' is equivalent to two conditionals, viz., 'If A is not B it is C,' and 'If A is not C it is B'; the latter of which is the converse of the former.

268. It is generally held by logicians that in a disjunctive not only does the truth of one member depend on the falsity of the other, but also the falsity of all but one depends on the truth of that one; so that, 'A is either B or C' would include, besides the two propositions given above, these two: 'If A is B it is not C', and its converse, 'If A is C it is not B'.

Of Complex Syllogisms.

269. A **Complex Syllogism** is one in which one or more complex propositions occur.

As complex propositions are of two kinds, Conditional and Disjunctive, so complex syllogisms will be of two kinds, conditional and disjunctive. If one premiss only is complex, it is called the major, and the simple premiss is the minor. If one premiss is conditional and the other disjunctive, the conditional premiss is called the major.

Of Conditional Syllogisms.

270. The most usual form of a conditional syllogism is where one premiss only (called the major) is conditional, the minor being simple. There are then two legitimate forms of reasoning. From the position, *i. e.* assertion, of the antecedent to the position, *i. e.* assertion, of the consequent; and, From the remotion, *i. e.* denial, of the consequent, to the remotion, *i. e.* denial, of the antecedent.

Thus, if the major is:

If A is B, C is D,

we may reason thus:

A is B; ∴ C is D.

Or,

C is not D; ∴ A is not B.

Ex. gr.:

If a man is shot through the heart he dies; X is shot through the heart, therefore he dies. Or, X did not die, therefore he was not shot through the heart.

271. It would be illegitimate to reason from the assertion of the consequent to the assertion of the antecedent, as: X died, therefore he was shot through the heart.

It is also illegitimate to reason from the denial of the antecedent to the denial of the consequent, as: X was not shot through the heart, therefore he will not die.

For it is not asserted that the consequent may not follow from other antecedents also.

272. If both premisses are conditional, the conclusion will be conditional. *Ex. gr.*:

If A is B, C is D:
If C is D, E is F:
∴ If A is B, E is F.

Of Disjunctive Syllogisms.

273. A **Disjunctive Syllogism** is one which has a disjunctive major and a simple minor. *Ex. gr.*:

Either A is B or C is D:
But A is not B; ∴ C is D.

If the disjunctive premiss has only two members, we may reason from the denial of either to the assertion of the others. If there are several members, we may reason from the denial of all but one to the assertion of that one, or from the denial of all but two or more to the assertion of these disjunctively. *Ex. gr.*: 'A successful man must have talents, industry, or good fortune'. 'So-and-so, who is successful, has neither talents nor industry; therefore he has good fortune'. Or, 'So-and-so has not talents; therefore he has either industry or good fortune'.

274. Logicians generally assume, as already stated, that the members of a disjunctive proposition are mutually exclusive, so that no two can be together asserted of the subject. If this is the case, we have a second valid form of reasoning, namely, from the assertion of one member to the denial of the rest. It follows, on this view, that if

we do not intend to exclude the possibility of two or more of the suppositions being true together, we must enumerate in the disjunctive all the possible combinations. 'A is either B (only) or C (only), or both B and C'. With three suppositions, as in the example above given, we should have seven members.

275. This view, however, does not agree with the ordinary use of language; and in the cases in which we do infer from the assertion of one member the denial of the rest, it is not from the form of the expression (unless the terms are opposed), but from our knowledge of the matter.

Of the Dilemma.

276. A **Dilemma** is defined as a syllogism which has a conditional major premiss, with more than one antecedent, and a disjunctive minor. *Ex. gr.*:

    If either A is B or C is D, X is Y:
    But either A is B or C is D:
∴ X is Y.

    If A is B, C is D; and if E is F, X is Y:
    But either A is B or E is F:
∴ Either C is D or X is Y.

In the latter case we might also reason thus:

    If A is B, C is D; and if E is F, X is Y:
    But either C is not D, or X is not Y:
∴ Either A is not B, or E is not F.

277. The following form is by some writers regarded as a dilemma, but is by others viewed as a conditional syllogism, the minor not being disjunctive.

If X is Y, either A is B or C is D:
But neither A is B nor C is D:
∴ X is not Y.

Here the minor simply denies the consequent of the major.

NOTE.—There is much difference of opinion amongst modern logicians as to the forms which should be included under the name Dilemma.

Of the Reduction of Complex Syllogisms.

278. It has already been observed that what a conditional proposition asserts is that the consequent follows from the antecedent.

If the conditional has only three terms, as: 'If A is B it is C'; then, as we have already seen, if these are common terms, it is reducible to 'Every AB is AC'. A syllogism with such a conditional premiss may therefore at once be reduced to a categorical form by substituting for the conditional premiss this equivalent; *ex. gr.:* 'If any bird devours grubs it is useful to the farmer. These birds devour grubs, ∴ they are useful to the farmer,' is reduced to, 'All birds which devour grubs are useful to the farmer. These birds devour grubs, ∴ they are useful to the farmer.' Similarly, if the conditional premiss is, 'If A is C, B is C, this is equivalent to, 'B is A'; and by substituting this equivalent, the syllogism is at once reduced to the categorical form. *Ex. gr.:* 'If birds that devour grubs are useful, then rooks are useful.' This is equivalent to ' Rooks devour grubs.'

279. If either subject is a singular term, the inference

may be from something not expressed, as: 'If Jones has written, the matter is settled'. 'If this quotation is found in Cicero at all, it must be in the philosophical treatises'.

In such cases we cannot reduce the reasoning to a categorical form without further information, simply because the actual grounds of the inference are not expressed.

280. Similarly, if there are four terms: 'If A is B, C is D'; there is no connexion between the antecedent and the consequent. The inference, therefore, depends on something not expressed.

281. Some logicians suggest reducing these by saying, 'The case of A being B is a case of C being D'. This is illusory. 'Case of' is ambiguous. It may mean 'instance of' (whether actual or imagined), and then this is only an awkward way of stating a categorical. 'Every case (instance) of A being B is a case of A being C' is equivalent to 'Every AB is AC'. 'If migratory birds breed in Ireland they winter farther south'. This may be stated, 'Every case of migratory birds breeding,' etc.; but this is exactly the same as, 'All migratory birds that breed in Ireland winter farther south'.

282. But 'case of' may mean 'hypothesis of', and then the proposition is still hypothetical, although disguised. *Ex. gr.*: 'If this witness is to be believed, the prisoner is innocent.' Here the prisoner's innocence is an individual fact; there are no 'cases,' *i.e.* 'instances,' of it. 'Case' here means 'hypothesis'; but the consequent is not a part of the antecedent, nor a species of it. The conclusion is really drawn from what the witness said, not from his credibility.

283. The hypothetical (= conditional) form is commonly adopted when we state a connexion of cause and effect, of

sign **and thing** signified, **or of** reason and indirect consequence, *i.e.* when steps intervene **which are** so obvious that they need not be stated. *Ex. gr.:* 'If the barometer falls there will be rain'. **Here** the relation is that of sign **and thing** signified, both being effects of the same cause.

CHAPTER IV.

Of Probable Reasoning.

§ 1.—*Of Chains of Probabilities.*

284. By **Probable Reasoning** is meant reasoning from propositions which are not certain. A proposition is usually called '**probable**' when it is more likely **to** be true than not; but the **word** '**probability**' **is** used to express any degree of likelihood or unlikelihood. When we speak of probable reasoning, we **use** the word probable in this wider sense. As almost all our reasoning is connected with probabilities, not certainties, it **is** important to ascertain what modification in **our** logical theory this fact involves. The chief points to be attended to **are** the following:

285. First, it is to be **observed that,** for convenience of calculation, **the** degree of probability of any proposition (or event) is represented by **a** fraction less than unity. The denominator is the whole number of possible cases: and **the** numerator **is** the number of those in which the proposition is true (or in which **the** event happens). Certainty is represented by unity; and as the proposition must

be either true or not true, the chance (or probability) of its falsehood plus the chance of its truth is equal to unity.

286. Now the chance that two (independent) statements are **both true together** is equal to the **product of the chances of their** truth separately.

Thus to take a pair of premisses in the first figure: S is M (let the chances be ¾): M is P (let the chances be ⅚). Then out of every 24 (= 4 × 6) S 18 are M, and out of these 18 M five-sixths, *i.e.* 15, are P. Hence out of 24 S we have 15 P; *i.e.* the chance of any given S being P is

$$\frac{3 \times 5}{4 \times 6}.*$$

287. It appears from this that in a chain of reasoning consisting of several propositions, each in itself only probable, the probability of the conclusion is diminished by every additional link. If the chain consisted of only three propositions, each having a chance = ⅘ (*i. e.* the odds in its favour being 4 to 1), the probability of the conclusion would be

$$= \frac{4 \times 4 \times 4}{5 \times 5 \times 5} = \frac{64}{125},$$

i. e. barely over ½.

§ 2.—*Of Cumulative Probabilities.*

288. This way of combining probable propositions in a chain must not be confounded with the accumulation of

---

* In an affirmative mood of the third figure the chance thus obtained would be the chance (not that any given S is P, but) that any given M is both S and P.

probable proofs each separately establishing the conclusion; as, for example,

>A is B (probably):
>B is C (probably):
>∴ A is C (with some degree of probability).

Also—

>A is D (probably):
>D is C (probably):
>∴ again, A is C (with some probability).

289. In a case like this, if either pair of premisses holds good, the conclusion **A is C** holds good; therefore, in order that **A** should not be **C**, both reasons must fail. Here then it is the falsity of the conclusion that involves a coincidence, the chance of which is the **product of** the separate chances of failure of the two grounds of proof. Suppose the chance that **A is C** by the first proof is $\frac{3}{4}$, and by the second $\frac{5}{6}$; then the chance of the first not holding good is $\frac{1}{4}$, that of the second failing is $\frac{1}{6}$; the product of these, $\frac{1}{24}$, is the chance that A is not C; therefore, so far, the chance that A is C = $\frac{23}{24}$.

290. If we had three reasons, each independently giving a chance = $\frac{4}{5}$, then the resulting chance of the conclusion would be

$$= 1 - \frac{1 \times 1 \times 1}{5 \times 5 \times 5} = 1 - \frac{1}{125} = \frac{124}{125}.$$

This is the **case of** what is called **Cumulative evidence or testimony.**

When probabilities have to be balanced, *i.e.* when there are reasons or evidence on both sides, the calculation is **more difficult.**

291. It deserves to be remarked that a conditional proposition, which expresses only a probable consequence, cannot be converted. From

If A is B, it is probably D,

we cannot infer,

If A is not D, it is probably not B.

For the latter is, as has been shown, the converse by contraposition of the former. But if the former is only probable it becomes, when put into a categorical form, I ('Some or Most B is D'), and I cannot be converted by contraposition. Even if it takes the form, 'Most B is D', we could not infer that there were amongst the not-D's any things that were not B. From 'Most University students are not honormen' we cannot infer that there are any honormen who are not University students.

Of Inductive Reasoning.

292. **Induction** proceeds from particulars to particulars, or from particulars to generals.

It assumes two principles:

First—**Whatever begins to exist has a cause.**

Second—Whenever the same circumstances occur the same result will occur, *i.e.* the **same causes always produce the same effects**, 'cause' being understood as including everything that may influence the effect.

293. This latter principle is the principle of **Uniformity of Nature**, and is the principle on which we act every time that we take food, or use any object, or in fact perform an action for any purpose whatever. The converse of this principle is obviously not true, since different causes may produce the same effect.

294. A special case of the principle of Uniformity of Nature is the principle of **Continuity**, viz., that when a certain relation is found to exist between the variations of the cause and those of the effect in several instances, the same relation, or one including it, exists in all the intermediate instances.

295. The following example illustrates the process of inductive reasoning:—Having read certain works of a novelist, and finding them witty and brilliant, we expect to find the same qualities in other works of the same author. In this inference we pass through an intermediate stage. Thus we first conclude that the author of these works possesses wit and talents, and then we infer from this, that his other works will display the same characters. That is to say, we reason first from the effects to the cause, and then from the cause to the effects.

296. So again, from the past mortality of all men we infer the mortality of men now living. Why? First, the mortality of men in all circumstances justifies us in concluding that it is the effect of something not peculiar to this or that man, place or circumstance, but belonging to human nature; and hence again we are justified in inferring that wherever human nature is found mortality belongs to it. This conclusion is further confirmed by other considerations; for example, similar experience in other animals, which enables us to say that the 'something' is not even peculiar to human nature, and again, the observation that death is only the climax of a series of changes always going on.

297. So also Newton, in establishing the law of gravitation, first ascended to the cause of the moon's revolution round the earth, *i. e.* the attractive force of the earth. He found, deductively, the law of this force. He then argued (from continuity) that a force which operates all through the moon's orbit at varying distances from the earth, according to a certain ratio, probably operates at other distances also, according to the same ratio, down even to the surface of the earth. It was easy to calculate what its amount at the surface would be, and this proved

to be exactly equal to the force which actually exists there, and which is, like the former, directed to the earth's centre. This coincidence was sufficient to prove the continuity of the law of attraction, at least to a high degree of probability. And this probability became certainty when it was found that the same law prevailed in other parts of space.

Of the Logical Basis of Induction.

298. The principles on which Induction rests are not themselves capable of proof, strictly so called. The principle of Uniformity of Nature makes a general statement as to the Unknown, viz., that in certain respects it resembles the Known; and this statement cannot be proved by either logic or experience, without taking for granted the very principle itself.

299. It is a question amongst logicians whether Induction can be brought under Deduction or not. If it is so, it must be by taking as a major premiss some form of the principle of Uniformity just stated. It comes to the same thing if we enumerate certain cases, and then assert that these are all the cases; *ex. gr.*, Mars, Jupiter, Venus, etc., revolve round the sun in ellipses. These are (= constitute) all the planets; ∴ All planets revolve round the sun in ellipses. In most cases we cannot enumerate all the instances; and the statement, 'These are all', means that for the purpose of the present argument these may be taken as representing all; *i.e.* that all the others may be supposed to resemble these. This involves the principle of Uniformity. The question, what condition must be fulfilled in order that this reasoning may be valid, does not belong to Formal but to Applied Logic.

Mr. Mill's View of the Type of Reasoning.

300. Mr. Mill considers that the true type of reasoning is that which proceeds from particulars to particulars, and, consequently, is not conclusive from the form of the ex-

pression. The universal proposition interposed as a major is, according to his view, not really a stage in the inference, but merely a convenient memorandum of past inferences, and a short formula for making more. When we say 'All men are mortal ; Jones is a man ; ∴ Jones is mortal', we do not infer the mortality of Jones from that of 'All men (including Jones'), but from the observed mortality of all men hitherto. The major, 'All men are mortal', means that we are justified in inferring from our past experience that any given man is mortal. Hence Mr. Mill regards syllogism as a process of interpretation (of our major), not as a process of inference. It is useful, therefore, as a test of the correctness of the reasoning by which we established the major.

301. This view may be applied to syllogisms in which the reasoning consists in bringing a particular case or group of cases under a general rule, *i.e.* to syllogisms which fall naturally into the first figure. It does not apply to syllogisms in the third figure with a definite middle.

Of Analogy.

302. **Analogy** is an argument in which from the resemblance of two things in certain respects we infer their resemblance in others; *ex. gr.*: if from the fact that Mars has many points of resemblance to the Earth we conclude that it is probably inhabited. It is from analogy we conclude that **vertebrate** animals (or even that other persons) feel pain as we do. Or the things compared may not be objects, but relations of objects. In this sense analogy is the resemblance of relations.

303. The term Analogy is, however, frequently used to signify imperfect induction, and by older writers to signify complete induction.

PART FOURTH.

CHAPTER I.

Of Fallacies.

304. A fallacy is an unsound argument. The unsoundness may be either—(1). In the reasoning considered in itself without regard to anything outside it; or, (2). In the reasoning considered in relation to a definite question proposed—as, for instance, the refutation of a given opinion.

305. The first class includes the fallacies in Expression. The second includes fallacies in the Matter.

306. The former class is again subdivided into fallacies which, in the form of the expression, violate the rules of logic, and fallacies which covertly violate them; that is to say, in which the violation appears only where the meaning of the terms is explained. These are called respectively Logical and Semi-logical fallacies. The latter are also called fallacies in dictione, *i. e.* in the wording.

Of Logical Fallacies.

307. Of strictly logical fallacies (otherwise called paralogisms) there might be as many species as there are

general rules of syllogism, which may be separately violated. There are, however, only three that deserve special enumeration here, viz., Illicit Process (taking a term more universally in the conclusion than in the premisses), Undistributed Middle, and Two Middle Terms.

308. **Illicit Process** may be either of the Major or of the Minor.

Of the major, as:

    Men of genius are (generally) eccentric:
    X is not eccentric:
    Therefore he is not a man of genius.

Even although the major should be given as probable (as in this case), we could not infer that the conclusion is probable. In order to ascertain whether this is so or not, we should know the proportion of not-eccentric persons who are men of genius, and this is not stated.

309. **Illicit Process of the Minor**, as:

    A, B, and C, are polite:
    A, B, and C, are Frenchmen:
    ∴ All Frenchmen are polite.

This is the fallacy in induction from an insufficient number of instances.

310. An example of **Undistributed Middle** is:

    All profound works are obscure:
    This is obscure:
    ∴ It is profound.

311. The fallacy of **Two Middle Terms** escapes notice

most readily when a verb is treated as if it were the copula, as:

    A resembles B :
    B resembles C :
   ∴ A resembles C.

A syllogism which is exactly the same in form as

    A is different from B :
    B is different from C :
   ∴ A is different from C.

Fallacies in Complex Syllogisms.

312. Under the head of logical fallacies are to be reckoned the two fallacious forms of inference from a conditional proposition, viz. from the assertion (position) of the consequent to the assertion of the antecedent, and from the denial (remotion) of the antecedent to the denial of the consequent.

313. A common instance of the former fallacy is, inferring the truth of the premisses, or the legitimacy of the reasoning, from the truth of the conclusion. It is often owing to a tacit inference of this kind that inconclusive arguments are brought forward by persons little likely to be deceived by a fallacy. They repeat an argument without examination, because they are firmly convinced of the truth of the conclusion.

314. They injure their cause by exposing it to the influence of the counter fallacy, which consists in inferring the falsehood of the conclusion from the falsehood of the premisses or the logical defect of the argument. This is

inferring the **denial of the consequent from** the denial of the antecedent. Even when good and bad arguments are mixed, there is a tendency to regard the bad as to a certain extent counterbalancing the good, instead of letting them go for nothing. An unfair disputant will of course attack the weak points only; and, when he has exposed these, will assume that he has refuted the conclusion.

In any particular syllogism these fallacies would resolve themselves into illicit process, undistributed middle, or two negative premisses.

Of Semi-logical Fallacies, or Fallacies in Dictione— Fallacy of Ambiguous Terms.

315. Of **Semi-logical Fallacies** the most important is that of **Ambiguous Terms**, called the Fallacy of **Equivocation**. The term most commonly ambiguous is the middle term, but either of the extremes may be ambiguous likewise. An example of ambiguous middle is:

> The old are more likely to be right in their judgment than the young:
> The men who wrote a thousand years ago are old writers:
> ∴ They are more likely to be right in their judgment than those of our own day.

Here the word 'old' is used in two senses—in the major, for those who have lived longer, and so have had more experience; and in the minor, for those who lived a long time ago, when the world was younger.*

> * 'Antiquitas sæculi juventus mundi'.

316. An important case of ambiguity is where a term is taken in one place collectively, and in another distributively. When we reason from the distributive sense to the collective, it is called the fallacy of **Composition**; when we reason from the collective to the distributive, it is the fallacy of **Division**. Overrating the probability of an inference from probable premisses comes under the former head. If the premisses are probably true (taken together), the conclusion is probably true: the premisses are probably true (each separately); ∴ the conclusion is probably true. Underrating the probability of an inference based on cumulative evidence comes under the head of the fallacy of division; as in the case of circumstantial evidence, where it is argued that this point and that point, and the third, etc., are not conclusive, *i. e.* separately, and it is inferred that they are not conclusive when taken together.

317. These two fallacies sometimes arise from the ambiguity of the words 'all,' 'some'. 'All' means either 'every' (distributively), or 'all together' (collectively). 'Some' has a somewhat similar ambiguity; sometimes meaning one or other, as when we say, 'Something must be done', 'You will meet some persons on the way'; and at other times, 'Some definite'.

318. Ambiguity may arise from a term being taken in one place in an abstract sense and in another in a concrete; *ex. gr.*,

Books are a solace to the weary:
Every book is either bound or unbound:
Therefore either bound books are a solace,
   or unbound books are so.

Again—

> Food is necessary to life :
> All food is either animal or vegetable :
> ∴ Either animal food is necessary, or vegetable food is necessary.

319. The Fallacy of **Accident** (Fallacia Accidentis), or *a dicto simpliciter ad dictum secundum quid*, consists in reasoning from a term stated without qualification to the same term with a qualification, as :

> You eat to-day what you bought in the market yesterday :
> What you bought was raw meat :
> Therefore you eat raw meat to day.

Here the term 'what you bought' is used in the first premiss of the substance only, but in the second of the substance in a particular state. If the first premiss had been stated with exactness, there would be palpably two middle terms.

320. The counter fallacy is called *a dicto secundum quid ad dictum simpliciter*, and consists in arguing from a statement with a particular qualification to a statement without the qualification, as :

> Opium is a poison
> Physicians give their patients opium :
> Therefore they give their patients poison.

321. There is an important ambiguity in words and expressions implying '**Sameness**' which deserves notice in a treatise on Logic, because sameness is sometimes that which is implied by the form of the proposition.

322. Strictly speaking, a thing cannot be the same with anything but itself. This sameness is for the sake of exactness sometimes called 'numerical identity,' *i. e.* there is only 'one' thing. When we speak of attributes, this meaning is modified. We call these the same when they are precisely similar, so that the same name applies to them. We speak of the same colour, sound, etc., meaning that the impressions produced on us are precisely similar. Similarly, when we say that two persons have the same disease, the same symptoms, etc., what we assert is similarity, sometimes including such similarity of origin as belongs to plants or animals which we say are of the 'same' species. We say, for instance, that one plant is the 'same' as another, meaning that the similarity is such as exists between plants produced from the seed of one identical plant. When we say that one man has the same organs as another, or as certain animals, what we predicate is precise similiarity of function, etc.

323. There is a still further extension of the notion when we speak of the skull as a modified vertebra, meaning that certain parts which we regard as essential to our conception of the plan of skull and vertebra may be described in the same terms. Still further, an organ is sometimes said to be a modification, development, survival, etc., of something which existed in an ancestor of the plant or animal. There is, of course, no one thing which has passed through the changes spoken of; but the plant or animal is treated as if it were one and the same with its own ancestors, and its parts numerically identical with the more or less similar parts in them, the only real unity being in the mind's conception of them. The discussion of this belongs rather to metaphysics than to logic, and it is only referred to here because relations of this kind are often disguised under the form of the simple proposition A is B.

Of Material Fallacies, or Fallacies extra Dictionem.

324. Of **Material Fallacies**, the first to be noted is **Ignoratio Elenchi**, *i.e.* irrelevant conclusion, or proving what is not the question. *Elenchus* means the refutation of an argument, and *ignoratio elenchi* strictly means ignoring the proper contradictory of the proposition to be refuted. This would include proving a particular, instead of a universal (*i.e.* proving a subcontrary instead of a contradictory), and proving an irrelevant conclusion, *ex. gr.* arguing that a thing is legally right, when the question is whether it is morally right; or, on the other hand, that it is either desirable or equitable, when the question is whether it is legal.

325. A third case comes under the definition, but can hardly be called a fallacy, viz. attempting to prove a universal when the proposition we wish to refute is universal (*i.e.* mistaking the contrary for the contradictory). This is not a fallacy, since the contrary includes the contradictory; but it exposes us to the risk of being refuted by the proof of a particular, and is therefore a serious fault in reasoning.

326. Petitio Principii, or Begging the Question, consists in taking as a premiss, without proof, a proposition which is equivalent to, or virtually involves, the conclusion; *ex. gr.* the ancient argument to prove that the earth is the centre of the universe. The point towards which all heavy bodies tend is the centre of the universe; but the point to which all heavy bodies tend is the centre of the earth; therefore the centre of the earth is the centre of the universe. Or again, the proof that a certain work was not written by its reputed author, since it contains allusions to a theory which (we assume) was not known in his day.

327. This fallacy is not limited to the cases in which the conclusion is logically contained in a single premiss. It must be remembered, that if one premiss of a syllogism, or all but one in a train of reasoning, be admitted, the truth of the conclusion turns on that of the remaining premiss. The fallacy consists in taking this for granted without attempt at proof. Inasmuch as in a legitimate syllogism the conclusion is contained in the premisses (taken together), the question may be asked, How does a legitimate syllogism differ from a *petitio principii*? The answer is, that in legitimate reasoning the premisses are either admitted or have been proved, whereas in the fallacy in question the very premiss which would be denied by those who reject the conclusion is taken for granted. But the actual syllogism is logically valid, the fallacy being in the matter, viz. in the assumption that the conclusion has been proved from admitted principles.

328. In **disjunctive reasoning** *petitio principii* very easily occurs, by the enumeration of alternatives in the disjunctive premiss being incomplete. Indeed, we can hardly ever be sure that it is complete unless the members are contradictory, and they may even appear to be contradictory when they are only contrary, as in the old dilemma to prove motion impossible. If a body moves, it moves either in the place where it is, or in the place where it is not; but both these are impossible, ∴ motion is impossible. By 'place where it is,' is meant the actual space it occupies. But then, 'in the place where it is' is not contradictory of 'in a place where it is not', both being sub-

divisions of 'in a place', which in the sense in which the word is taken is contradictory of motion. The alternative omitted, namely, 'from the place where it is to the place where it is not', is the only possible one.*

329. The question may be begged by a single word; for instance, if we speak of an opinion we are attacking as 'foolish', 'heretical', 'audacious', etc. These have been called 'question-begging epithets', since they connote the very thing to be proved.

330. **Arguing in a circle** may be regarded as a species of petitio principii. It consists in using the conclusion to prove the disputed premiss, from which again the conclusion is inferred. For instance, in the example above given: 'This work is not the production of the reputed author, since it contains allusions to a theory not known in his time'; if we are challenged to prove the assumption, and attempt to do so by saying, 'It is not referred to in any genuine work of that age', we now assume that the work in question is not genuine. This fallacy is likely to be committed when we attempt to prove a principle which is incapable of strict proof, only because it is above proof. For example, an attempt is sometimes made to prove from experience the principle used in induction, viz. that in like circumstances like results follow. Thus: We have hitherto found that in like circumstances like results have followed; therefore we are justified (*a*) in concluding that in like circumstances like results will always be found. Here we already assume at (*a*) the very principle which we profess

---

* The alternative, 'partly in one place, and partly in the other', is inadmissible. A part of a body cannot any more than the whole be in two places at once.

to prove in the conclusion, viz. that we can argue from the known to the unknown.

331. The fallacy called **a non causa pro causa** consists in inferring a certain effect from something which is not really a cause of it; *ex. gr.* :

There will be war, for a comet has appeared.
There will be a change in the weather, for there has just been a change in the moon.

A common case of this fallacy is the assumption that one thing is the effect of another, merely because it has followed it; for instance, that an increase or decrease in the prosperity of a nation is the effect of some particular measure, or that rain takes the cold out of the air. This form of the fallacy is called the fallacy of reasoning, **Post hoc ergo propter hoc.**

332. Another common case is, inferring that one thing is the effect of another, because it has often followed it, without regarding the instances in which it has not followed it. It is thus that prophetic dreams, &c., which appear to come to pass, are carefully noted, while those which fail are forgotten.

333. The fallacy of **Many Questions** (*plurium interrogationum*) consists in combining two or more questions in one, and insisting on a simple answer. A traditional example is: Have you left off beating your father?

334. The fallacy of **False Analogy** consists in inferring from a resemblance of relations a resemblance of the things themselves in other relations. This is often aided by the metaphorical use of language. Thus a nation or a city is spoken of as feminine, and depicted in the form of a woman, solely because it is conceived as giving birth

to its 'sons'; but the figure has been used as **if a nation resembled a** woman in feebleness and helplessness.

335. The **Argumentum ad hominem** is an argument founded on premises not supposed to be universally admitted, but admitted **by** a particular opponent. The inference then **is not** absolute, that **the** conclusion is true, **but** conditional—that the person who accepts the premises must accept the conclusion. It becomes a fallacy **only** when this conditional conclusion **is** assumed **to be** absolute **or** categorical.

CHAPTER II.

Of Methods of Proof and Exposition.

336. The **order of Exposition** of a system of knowledge may be either **Analytic** or **Synthetic.**

337. **The Analytic** method proceeds from the complex to the simple; from results to principles. Thus in Logic **the** analytic method would commence **with** chains of reasoning, and proceed through syllogisms **and** propositions to terms. Or, in Astronomy, again, it would start from the apparent motions of the planets, and proceed to the fact of their revolution round the sun, arriving, **as a** last result, at the law of gravitation.

338. The **Synthetic** method proceeds from the simple to the complex; **from** principles **to results.** Such is, in Logic, the method usually adopted, which commences with Terms and ends with complex reasonings; and in Astronomy, **the** method which begins with the law of

gravitation and deduces from it the apparent motions of the planets. The Analytic method is suited to Discovery; the Synthetic is suited to Instruction.

339. **Methods of Proof** are either **a priori** or **a posteriori**.

The **a priori** proof is drawn from that which is logically first, viz., from the cause or the general law.

The **a posteriori** from the effect or the actual observation of facts. Thus the difference in the rate of vibration of pendulums of different lengths is proved a priori when it is deduced from mechanical principles, and a posteriori when proved by actual experiment.

340. **Explanation** consists in referring a particular phenomenon to a general law. Thus we are said to explain the fall of a stone when we show that it is a consequence of the general law of gravitation. By 'consequence' is here meant logical consequence. The law of gravitation is not the cause of the stone falling. Gravitation may be called the cause, but is really only a general name for a large class of similar facts.

341. It is clear that there is a limit to explanation. We at last reach the most general principle—one which is incapable of explanation. This does not mean that it is more mysterious than other facts, but that it is more general, and that we know none more general. Isolated facts are sometimes called mysterious when we are not able to bring them under any law, and yet, as there is no isolated fact in nature, we believe that there is some law in the case.

342. Explanation is Analytic : Deduction is Synthetic.

The fewer the general principles which we require to assume, the more perfect in form is our science.

343. But it is clear, as was said of explanation, that there must be some limit. There must be some principles in every science or subject which are incapable of proof; just as a chain cannot hang unsupported. Thus, the principles already referred to, of Universal Causation, and of the Uniformity of Nature, cannot be logically proved. Yet it would be an error to suppose that they rest on a weaker foundation than other familiar laws of nature, which we consider as proved, for the latter ultimately depend on these and other unproved principles, and the inference cannot be more certain than the principle from which it is inferred.

APPENDIX.

EXERCISES.

In stating a proposition in strictly logical form, whether for the purpose of immediate inference or of syllogism, the student must bear in mind—

(1). That when the predicate is a verb it must be resolved so that the copula shall be simply 'is' or 'is not.'

(2). That the grammatical subject is not always the real or logical subject; *ex. gr.*, 'There was no decisive result from this experiment', or, 'No inference can be drawn from this experiment'. Here the real subject is, 'This experiment', of which it is asserted that it had no decisive result.

In order to ascertain what the real subject and predicate are, consider what it is that an assertion is made about, and what it is that is asserted.

(3). When any proposition is compound (copulative, adversative, etc.) it must be resolved.

(4). Some of the arguments in the following exercises are valid, and some invalid. Having brought the reasoning into apparently syllogistic form, then, if this is categorical, the student should examine—(*a*) whether there are more than three terms (really and not merely in appearance); (*b*) whether the middle term is distributed; (*c*) whether any extreme which is distributed in the conclusion has been distributed in the premisses. If there is a disjunctive major, care must be taken that the enumeration of alternatives is complete. Of course the other rules of correct reasoning must not be overlooked, but these are the points most likely to escape notice.

Propositions.

State the following propositions in strictly logical form, stating, in the case of simple categorical propositions, what is the subject, what the predicate, and what the quantity and quality of the proposition; and, in the case of complex propositions, whether they are conditional or disjunctive :—

1. Troja fuit.
2. Humanum est errare.
3. Much study is a weariness to the flesh.
4. All is not gold that glitters.
5. Many a little makes a mickle.
6. Sapientis est providere.
7. Great is Diana of the Ephesians.
8. All is not truth that is confidently asserted.
9. The most honest statesmen are not always the most popular.
10. Scholarship is not what it was.
11. Poeta nascitur, non fit.
12. A little learning is a dangerous thing.
13. 'Books are not absolutely dead'. (Milton) (wrongly resolved by Mr. Jevons into 'Some books are living'.)
14. That may be legal, but it is not equitable.
15. Things equal to the same are equal to each other. (Observe that in this proposition we could not substitute 'every thing' for 'things' in the subject.)
16. It was Newton that discovered the law of gravitation.
17. It was not Newton that discovered the orbits of the planets.
18. Whales are not the only marine mammalia.
19. All these duties are too much for me.
20. Few know how little they know.

21. 'God did not make men barely two-legged creatures, and leave it to Aristotle to make them rational' (Locke).
22. To be or not to be, that is the question'.
23. 'On earth there is nothing great but man; In man there is nothing great but mind'.
24. 'Non tali auxilio nec defensoribus illis Tempus eget'.
25. There is something unreasonable in most men.
26. Truth is stranger than fiction.
27. Such liberty is only licence.
28. Gold is the monetary standard of Great Britain. (We cannot assert either of 'all gold' or of 'some gold' that it is the standard.)
29. Hops are the staple produce of Kent.
30. No man is wholly bad.
31. 'An honest man's the noblest work of God'.
32. 'Adsum qui feci'.
33. After all, a promise is a promise. (How would you contradict this?)
34. The wisest of men sometimes errs.
35. Books are a source both of instruction and amusement.

Conversion.

Convert the following propositions:—

1. A is father of B.
2. Cain killed Abel.
3. Jones struck Smith.
4. All cats have been kittens.
5. All water contains air.
6. It rains. (Incorrectly converted by Jevons thus: Something that is letting rain fall is the atmosphere. Test this by converting the following proposition, which is similar.)
7. It is freezing. (N.B.—These must be regarded as propositions expressing Real Existence.)
8. To be righteous is to be happy.
9. Boys will be boys.

APPENDIX.

10. Judges ought to be impartial.
11. Life all men hold dear.

Related Propositions.

Assign the logical relation between the propositions in each of the following groups :—

1.
    Every A is B.
    No A is B.
    Some Bs are not A.
    Some As are not B.
    Some things not A are not B.
    Some Bs are A.
    Some things not A are not B.
    Every B is A.
    Nothing not-A is B.

2. All birds are bipeds.
    Some things not birds are bipeds.
    Some things not bipeds are birds.
    Some bipeds are not birds.
    No birds are quadrupeds.
    No birds are not bipeds.
    Bipeds alone are birds.

3. I believe this story: I disbelieve this story.
4. I ought to do this: I ought not to do this.
5. The Tories are always right: The Tories are always wrong.
6. Is either of the following propositions true, and if so, which ?—

> All Englishmen who do not take snuff are to be found among Europeans who do not use tobacco.
> All Englishmen who do not use tobacco are to be found among Europeans who do not take snuff (De Morgan).

APPENDIX

Immediate Inferences.

Examine the following immediate inferences: if they are correct, state the logical principle under which they come; if incorrect, point out the fallacy:—

1. Every old man has been a boy; ∴ Some boys will be old men.
2. No animals are self-made = All animals are not self-made; ∴ All self-made things are not animals; ∴ Some things not animals are self-made.
3. No persons of the male sex are winged; ∴ No winged persons are of the male sex; ∴ Some winged persons are not of the male sex; ∴ Some persons not of the male sex are winged.
4. A man is an animal; ∴ The head of a man is the head of an animal.
5. Truth always triumphs; ∴ Whatever opinion has triumphed is true.

On the Moods.

1. Show that if we substitute the conclusion for the major premiss in Barbara or Celarent we obtain legitimate premisses in the third figure, giving as a conclusion the subalterna of the original major.
2. If in Bramantip we substitute the conclusion for the minor premiss we obtain premisses which in the first figure give a conclusion, the converse per accidens of the original minor, or in the fourth figure would give a conclusion the subalterna of the original minor.
3. If in Cesare we substitute the conclusion for the major, we can draw in the third figure and mood Felapton, a conclusion which is the subalterna of the converse of the original major.
4. If in Camenes we substitute the conclusion for the minor, we obtain premisses which will give in the fourth figure and mood Fesapo a conclusion, the subalterna of the original minor.
5. If in Camestres we substitute the conclusion for the minor premiss, we obtain premisses which in the fourth

figure and mood Fesapo lead to a conclusion, the subalterna of the converse of the original minor.

6. Show that in no other case will the substitution of the conclusion for a premiss furnish a legitimate pair of premisses.

7. Show that in the third figure we may with the same minor premiss have contradictory major premisses, and that the conclusions will be subcontrary.

8. Show that the same is true of the fourth figure.

9. Show that in no other case can any conclusion be drawn from one premiss and the contradictory of the other.

Miscellaneous Examples of Arguments.

Reduce the following reasonings to strict logical form, if possible: if the syllogism is simple, find mood and figure; if complex, find to what species of complex reasoning it belongs. If the reasoning is invalid, show where the fallacy lies:—

1. None but whites are civilized: the ancient Germans were whites; ∴ they were civilized.

2. None but whites are civilized: the Hindoos are not whites; ∴ they are not civilized.

3. None but civilized people are whites: the Gauls were whites; ∴ they were civilized.

4. This disease is not infectious; for A and B were exposed to it, and did not take it.

5. All moral precepts are binding on every man: some of the precepts of Confucius are moral; ∴ the precepts of Confucius are, to a certain extent, binding on every man.

6. Some vertebrates are bipeds: some bipeds are birds; ∴ some birds are vertebrates.

7. Whales are not true fishes, for they cannot breathe in water, and, besides, they suckle their young.

8. Food is necessary to life: whatever is necessary to life must exist in all inhabited countries; ∴ there are certain kinds of food that exist in all inhabited countries.

9. Every B is A: only C is A; ∴ only C is B.

10. Snow is white: white is a colour; ∴ Snow is a colour.

11. The sun is insensible: the Persians worship the sun; ∴ the Persians worship a thing insensible (Port Royal Logic).

12. That which does not consist of parts cannot perish by the dissolution of its parts: the soul has no parts; ∴ the soul cannot perish by the dissolution of its parts. (There seem to be two negative premisses; is this really the case?) (Port Royal Logic.)

13. He who believes himself to be always right in his opinion lays claim to infallibility: you always believe yourself to be in the right in your opinion (else it would not be your opinion); ∴ you lay claim to infallibility. (Whately).

14. If benevolence were the whole of virtue, we should not approve of benefits done to one more than to another, as we actually do, inasmuch as we approve of gratitude and acts of friendship. (Butler.)

15. Our notion of man includes a certain figure, as well as rationality, for if a parrot talked rationally we should not call it a man. (Locke.)

16. If the germination called spontaneous did not depend on external germs, it would occur in perfectly pure air; but it does not. (The conclusion is suppressed.) (Try to reduce the reasoning to a categorical form.)

17. If the planets shone by their own light, Venus would not show phases. (The conclusion and one premiss are suppressed.)

18. If the Claimant were the person he pretended to be he would not have forgotten his Virgil so completely.

19. If education alone made men moral, the Redpaths and Robsons would not have been guilty of forgery. (More than one syllogism is implied.)

20. A historian ought to be impartial: in order to be impartial it is necessary to know what has been said on both sides; ∴ a historian ought to know what has been said on both sides.

21. Every good book is worth reading more than once: few books are worth reading more than once; ∴ few books are good books. (This seems to be valid reasoning, and

yet it seems to be a syllogism in the second figure with both premisses affirmative.)

22. No evil should be allowed that good may come of it: punishment is an evil; ∴ punishment should not be allowed that good may come of it. (Whately.)

23. If, as you say, every student ought to read this book, it would probably sell well; but it does not, therefore some students at least ought not to read it.

24. Unless some one else has locked the door, I must lock it; I find some one has done so; therefore I must not.

25. Vaccination is no protection whatever against smallpox; for A, B, and C were vaccinated, and yet have taken smallpox.

26. Exposure to cold is good for children, for all the grown people who have been exposed to cold as children are strong.

27. No soldiers should be brought into the field who are not well qualified to perform their parts: none but veterans are well qualified to perform their parts; therefore none but veterans should be brought into the field.

28. Truth always triumphs: the present theory has triumphed; ∴ it is true.

29. Truth always triumphs over persecution: this doctrine has not triumphed over persecution (never having been persecuted); ∴ it is not true.

30. No X is Y: no Z is X; ∴ some things not Y are not Z.

31. No judge is infallible; no judge is unskilled in law; ∴ some persons skilled in law are infallible.

32. If competitive examinations do not tend to the selection of the best men they ought to be abolished; if they do so tend they should be applied to every appointment; therefore either competitive examinations should be abolished, or they should be applied to every appointment.

33. 'The essences of the species of things are nothing else but abstract ideas. For the having the essence of any species being that which makes anything to be of that species, and the conformity to the idea to which the name is annexed being that which gives a right to that name, the having the essence and the having that conformity

APPENDIX

must needs be the same thing; since to be of any species and to have a right to the name of that species is all one.' (Locke.)

34. The perfection of virtue consists in perfect harmony with right, so that right is done without a struggle, and the nearer a man approaches to this, the more perfect his virtue is; ∴ the greater the virtue, the less the self-denial.

35. The stronger the temptations and the inclination to wrong-doing, the greater is the virtue which, in spite of these, adheres to the right; ∴ the greater the self-denial the greater the virtue.

36. 'If there be no difference between inward principles but that of strength, we can make no distinction between the murder of a father and an act of filial duty; ... but in our coolest moments must approve or disapprove them equally: than which nothing can be reduced to a greater absurdity.' (Butler.)

37. Men in some ages have thought usury a vice; men in other ages have not; ∴ the standards of virtue and vice are not invariable.

Exercises on Probability.

1. An urn contains 1000 balls, numbered 1 to 1000, from which one is drawn. A witness, whose average credibility $= \rho$, testifies that a particular number, say 926, has been drawn. What is the credibility of this particular assertion?

We must make some assumption as to the witness's knowledge of the number of balls in the urn. Let us, then, assume that he knows the number. Then in the case given there are two possible hypotheses:—

First Hypothesis—926 has been drawn.

Here two things coincide. No. 926 is drawn; the chance of this being $\frac{1}{1000}$, and the witness speaks truth, the chance of this being ρ. Hence the chance of the coincidence is proportional to the product, $\frac{\rho}{1000}$.

Second Hypothesis—926 has not been drawn.

Here three things coincide. No. 926 not drawn; the chance of this is $\frac{999}{1000}$; the witness speaks falsely; the chance of which $= 1 - \rho$, and thirdly, out of the 999 balls not drawn, he names this particular number; the **chance** of his doing so being $\frac{1}{999}$. The chance of the **coincidence is** proportional **to the** product of these three, *i. e.* to $\frac{1-\rho}{1000}$. Therefore the chances in **favour of** his evidence being true are to **those** against as ρ to $1-\rho$. **The** resulting probability is therefore ρ, that **is to say, the same as the witness's average credibility.**

2. **Suppose the witness not** to know the number of **balls.**

3. **An** urn **contains** 1 black, 99 red, and 900 white balls. One is drawn, and **a witness,** whose average credibility $= \rho$, announces that black has been **drawn.** What is the degree of credibility of his statement?

The data are insufficient. We must be **told** what is the number of possible false assertions, that is, in this case, **how** many different colours the witness is liable to mention falsely, whether by mistake or otherwise. Let us first assume that he has the choice of 13 in all.

First Hypothesis—

    Black is drawn, $= \frac{1}{1000}$,

    The witness speaks **truly,** $= \rho$.

        Product, $= \frac{\rho}{1000}$

Second Hypothesis—

    Black was not **drawn,** $= \frac{999}{1000}$,

    The witness speaks falsely, $= 1-\rho$.

    Of the twelve colours not drawn, he specifies black, $= \frac{1}{12}$.

        Product, $= \frac{999(1-\rho)}{12000}$.

Here the **chances** in favour **of** his truthfulness are to those against as 12ρ to $999(1-\rho)$. The probability may therefore be stated as $\frac{12\rho}{12\rho + 99(1-\rho)}$. Had he **announced white,** the probability would have been $\frac{108\rho}{108\rho + (1-\rho)} = \frac{108\rho}{107\rho + 1}$.

4. An event whose antecedent probability is v is announced by a witness whose average credibility is ρ. Let the antecedent probability of his announcing this event falsely be x. What are the chances that his statement is true?

INDEX.

The numbers refer to the Paragraphs.

Abstraction, 19
Abstract Terms, 9.
Accident, Fallacy of, 319.
Analogy, 302.
———— False, 334.
Analytical Propositions, 96.
Analytic Method, 336.
A posteriori Proof, 339.
A priori **Proof,** 339.
Argumentum ad hominem, 335.
Aristotle's Dictum, 231.

Begging the question, 326.

Categories, 111.
Categorical, 164.
Circle, arguing in, 330.
Clearness, 35.
Common Terms, 12.
Complex Propositions, 261.
———— Syllogisms, 269.
Comprehension, 16.
Concepts, **25.**
Concrete Terms, 9.
Conditional Propositions, **262.**
———— Syllogisms, 270.
Conjunctive Propositions ; the name given by some logicians to **Conditional** Propositions.

Connotation, 16.
Continuity, Law of, 294.
Contradiction, Law of, 31.
Contradictory Propositions, **120.**
Contraposition, 148.
Contrary Terms, 33.
———— Propositions, **126.**
Conversion, **135.**
'Converse,' incorrect use of, **150.**
Copula, 56.
Copulative Propositions, 100.
Cumulative Probabilities, 288.

Definition, 38.
Denotation, 16.
Dilemma, **276.**
Disjunctive Propositions, **266.**
———— Syllogisms, 273.
———— Fallacies in, 328.
Distinctness, 35.
Distribution, 73.
Division, 47.

Enthymeme, 256.
Equipollent, 62.
Excluded Middle, Law of, 32.
Explanation, 340.
Extension, 16.

INDEX.

Fallacies, 304.
Figure, 178.
Figures, Uses of, 209.
Form, 2.

Genus, 20.
────── Summum, 24.

Higher Notions, 22.
Hypothetical Propositions, 262.

Ignoratio Elenchi, 324.
Illicit process, 169.
Induction, 292.
Inverse: 'Every A is B' is called the inverse of 'Every B is A.'

Matter, 2.
Mill's view of Syllogism, 300.
Modal Propositions, 104.
Moods, 213.

Non-connotative Terms, 27.

Obverse: the same as 'equipollent': see 62.
Obversion: the change of a proposition into its equipollent; by some logicians called Permutation.
Opposition, 119.

Paralogism, 307.
Permutation: see Obversion.
Petitio principii, 326.
Post hoc, Ergo propter hoc, 331.
Predicables, 107.

Predicament = 'Category'. 'Predicamentum' was formed by Latin logicians in imitation of the Greek Κατηγορία.
Probable Reasoning, 284.
Property, 109.
Propositions, Import of, 88.

Quality, 60.
Quantity, 60.
Quantified, 70.
Quantification of Predicate, 82.

Real Propositions, 96.
Reduction, 236.
────────── of Complex Syllogisms, 278.
Reductio ad Impossibile, 250.

'Same,' ambiguity of, 321.
Semi-logical fallacies, 306, 315.
Singular Terms, 12.
Sorites, 257.
Species, 20.
────── infima, 24.
Subalternation, 116.
Subcontrariety, 130.
Substitution, Principle of, 157.
Syllogism, 162.
Synthetical, Propositions, 96.
────────── Method, 336.

Term, 7.

Unfigured Syllogism, 253.
Uniformity of Nature, Law of, 293.

Verbal Propositions, 96.

THE END.

www.ingramcontent.com/pod-product-compliance
Lightning Source LLC
Chambersburg PA
CBHW020139170426
43199CB00010B/813